British and American
Festivities

英美節慶趣話

商務印書館

This Chinese edition of *British and American Festivities*
has been published with the written permission of
Black Cat Publishing.

The copyright of this Chinese edition is owned by
The Commercial Press (H.K.) Ltd.

Name of Book: British and American Festivities
Author: Gina D. B. Clemen
Editors: Monika Marszewska
Design: Nadia Maestri
Edition: ©1999 Black Cat Publishing
                 an imprint of Cideb Editrice, Genoa, Canterbury

系 列 名：Black Cat 優質英語階梯閱讀 · Level 2
書 　 名：英美節慶趣話
責任編輯：傅　伊
封面設計：張　毅
出 　 版：商務印書館 (香港) 有限公司
　　　　　香港筲箕灣耀興道 3 號東滙廣場 8 樓
　　　　　http://www.commercialpress.com.hk
發 　 行：香港聯合書刊物流有限公司
　　　　　香港新界大埔汀麗路 36 號中華商務印刷大廈 3 字樓
印 　 刷：中華商務彩色印刷有限公司
　　　　　香港新界大埔汀麗路 36 號中華商務印刷大廈
版 　 次：2015 年 11 月第 8 次印刷
　　　　　© 2003 商務印書館 (香港) 有限公司
　　　　　ISBN 978 962 07 1638 6
　　　　　Printed in Hong Kong

出版説明

　　本館一向倡導優質閲讀，近年來連續推出了以"Q"為標識的 "Quality English Learning 優質英語學習"系列，其中《讀名著學英語》叢書，更是香港書展入選好書，讀者反響令人鼓舞。推動社會閲讀風氣，推動英語經典閲讀，藉閲讀拓廣世界視野，提高英語水平，已經成為一種潮流。

　　然良好閲讀習慣的養成非一日之功，大多數初、中級程度的讀者，常視直接閲讀厚重的原著為畏途。如何給年輕的讀者提供切實的指引和幫助，如何既提供優質的學習素材，又提供名師的教學方法，是當下社會關注的重要問題。針對這種情況，本館特別延請香港名校名師，根據多年豐富的教學經驗，精選海外適合初、中級英語程度讀者的優質經典讀物，有系統地出版了這套叢書，名為《Black Cat 優質英語階梯閲讀》。

　　《Black Cat 優質英語階梯閲讀》體現了香港名校名師堅持經典學習的教學理念，以及多年行之有效的學習方法。既有經過改寫和縮寫的經典名著，又有富創意的現代作品；既有精心設計的聽、說、讀、寫綜合練習，又有豐富的歷史文化知識；既有彩色插圖、繪圖和照片，又有英美專業演員朗讀作品的 CD。適合口味不同的讀者享受閲讀之樂，欣賞經典之美。

　　《Black Cat 優質英語階梯閲讀》由淺入深，逐階提升，好像參與一個尋寶遊戲，入門並不難，但要真正尋得寶藏，需要投入，更需要堅持。只有置身其中的人，才能體味純正英語的魅力，領略得到真寶的快樂。當英語閲讀成為自己生活的一部分，英語水平的提高自然水到渠成。

商務印書館 (香港) 有限公司
編輯部

使用說明

1 應該怎樣選書？

按閱讀興趣選書

《Black Cat 優質英語階梯閱讀》精選世界經典作品，也包括富於創意的現代作品；既有膾炙人口的小說、戲劇，又有非小說類的文化知識讀物，品種豐富，內容多樣，適合口味不同的讀者挑選自己感興趣的書，享受閱讀的樂趣。

按英語程度選書

《Black Cat 優質英語階梯閱讀》現設 Level 1 至 Level 6，由淺入深，涵蓋初、中級英語程度。讀物分級採用了國際上通用的劃分標準，主要以詞彙（vocabulary）和結構（structures）劃分。

Level 1 至 Level 3 出現的詞彙較淺顯，相對深的核心詞彙均配上中文解釋，節省讀者查找詞典的時間，以專心理解正文內容。在註釋的幫助下，讀者若能流暢地閱讀正文內容，就不用擔心這一本書程度過深。

Level 1 至 Level 3 出現的動詞時態形式和句子結構比較簡單。動詞時態形式以現在時（present simple）、現在時進行式（present continuous）、過去時（past simple）為主，句子結構大部分是簡單句（simple sentences）。此外，還包括比較級和最高級（comparative and superlative forms）、可數和不可數名詞（countable and uncountable nouns）以及冠詞（articles）等語法知識點。

Level 4 至 Level 6 出現的動詞時態形式，以現在完成時（present perfect）、現在完成時進行式（present perfect continuous）、過去完成時（past perfect continuous）為主，句子結構大部分是複合句（compound sentences）、條件從句（1st and 2nd conditional sentences）等。此外，還包括情態動詞（modal verbs）、被動形式（passive forms）、動名詞（gerunds）、

短語動詞（phrasal verbs）等語法知識點。

根據上述的語法範圍，讀者可按自己實際的英語水平，如詞彙量、語法知識、理解能力、閱讀能力等自主選擇，不再受制於學校年級劃分或學歷高低的約束，完全根據個人需要選擇合適的讀物。

② 怎樣提高閱讀效果？

閱讀的方法主要有兩種：一是泛讀，二是精讀。兩者各有功能，適當地結合使用，相輔相成，有事半功倍之效。

泛讀，指閱讀大量適合自己程度（可稍淺，但不能過深）、不同內容、風格、體裁的讀物，但求明白內容大意，不用花費太多時間鑽研細節，主要作用是多接觸英語，減輕對它的生疏感，鞏固以前所學過的英語，讓腦子在潛意識中吸收詞彙用法、語法結構等。

精讀，指小心認真地閱讀內容精彩、組織有條理、遣詞造句又正確的作品，着重點在於理解"準確"及"深入"，欣賞其精彩獨到之處。精讀時，可充分利用書中精心設計的練習，學習掌握有用的英語詞彙和語法知識。精讀後，可再花十分鐘朗讀其中一小段有趣的文字，邊唸邊細心領會文字的結構和意思。

《Black Cat 優質英語階梯閱讀》中的作品均值得精讀，如時間有限，不妨嘗試每兩個星期泛讀一本，輔以每星期挑選書中一章精彩的文字精讀。要學好英語，持之以恆地泛讀和精讀英文是最有效的方法。

③ 本系列的練習與測試有何功能？

《Black Cat 優質英語階梯閱讀》特別注重練習的設計，為讀者考慮周到，切合實用需求，學習功能強。每章後均配有訓練聽、說、讀、寫四項技能的練習，分量、難度恰到好處。

聽力練習分兩類，一是重聽故事回答問題，二是聆聽主角對話、書信朗讀、或模擬記者訪問後寫出答案，旨在以生活化的練習形式逐步提高聽力。每本書均配有CD提供作品朗讀，朗讀者都是專業演員，英國作品由英國演員錄音，美國作品由美國演員錄音，務求增加聆聽的真實感和感染力。多聆聽英式和美式英語兩種發音，可讓讀者熟悉二者的差異，逐漸培養分辨英美發音的能力，提高聆聽理解的準確度。此外，模仿錄音朗讀故事或模仿主人翁在戲劇中的對白，都是訓練口語能力的好方法。

閱讀理解練習形式多樣化，有縱橫字謎、配對、填空、字句重組等等，注重訓練讀者的理解、推敲和聯想等多種閱讀技能。

寫作練習尤具新意，教讀者使用網式圖示（spidergrams）記錄重點，採用問答、書信、電報、記者採訪等多樣化形式，鼓勵讀者動手寫作。

書後更設有升級測試（Exit Test）及答案，供讀者檢查學習效果。充分利用書中的練習和測試，可全面提升聽、說、讀、寫四項技能。

❹ 本系列還能提供甚麼幫助？

《Black Cat 優質英語階梯閱讀》提倡豐富多元的現代閱讀，巧用書中提供的資訊，有助於提升英語理解力，擴闊視野。

每本書都設有專章介紹相關的歷史文化知識，經典名著更有作者生平、社會背景等資訊。書內富有表現力的彩色插圖、繪圖和照片，使閱讀充滿趣味，部分加上如何解讀古典名畫的指導，增長見識。有的書還提供一些與主題相關的網址，比如關於不同國家的節慶源流的網址，讓讀者多利用網上資源增進知識。

Contents

The story is recorded in full. 故事錄音

This symbol indicates the exercises featured on the accompanying CD.
聽力練習的錄音標記

Introduction

How much do you know about British and American festivities [1]? Do you know that ancient [2] pagan customs [3] are still part of some festivities? This book tells you about the most important festivities of the year, their origins [4] and how they are celebrated.

You will notice that some festivities are celebrated only in Great Britain and others only in America. You will also notice that Americans love big celebrations with parades [5] and marching bands [6].

If you want to know more about a festivity, you can contact the Internet sites listed on page 116.

---

1. **festivities** : 節日。
2. **ancient** : 古老的。
3. **pagan customs** : 異教風俗。
4. **origins** : 起源。
5. **parades** : 遊行。
6. **marching bands** : 排成方隊行進的樂隊。

CHAPTER ONE

Columbus Day

he second Monday of October is Columbus Day! This festivity is celebrated only in the United States of America. Why?

During Christopher Columbus' time people thought the world was flat. Columbus was born in Genoa, a beautiful Italian city on the north-west coast. For many years he watched ships leave the port [1] of Genoa. He noticed that these ships seemed to go under the horizon [2]. He was convinced [3] that the world was round, but no one believed him. He wanted to reach the East by sailing to the West. Queen Isabella and King Ferdinand of Spain believed Columbus' theory. They gave him three ships, the *Santa Maria*, the *Niña* and the *Pinta* to travel and test this theory.

Columbus was an expert navigator [4]. After a long and difficult voyage, he and his men reached the North American continent [5] on 12 October, 1492! The famous Italian explorer discovered the New World! After his great discovery a new era [6] of exploration [7] began and America was born.

1. **port** : 港口城鎮。
2. **horizon** : 地平線。
3. **was convinced** : 確信。
4. **navigator** : 導航者。
5. **continent** : 大陸。
6. **era** : 時代。
7. **exploration** : 考察。

Columbus Day

Today most Americans celebrate Columbus Day with colourful parades and they elect [1] a Columbus Day Queen. The parade is usually long with big floats [2] dedicated to [3] Columbus and there are other multi-cultural [4] floats too. After the parade there is usually a dinner and dance.

In San Francisco, California, there is also a re-enactment [5] of Columbus' discovery. A man dresses up [6] as Columbus and several other men dress up as his sailors. They get into a boat and row to the beach. When they get out of their boat they kneel on the beach and thank God.

Americans enjoy remembering the great navigator!

A Columbus Day float in New York.

1. **elect**：選舉。
2. **floats**：彩車。
3. **dedicated to**：紀念。
4. **multi-cultural**：多種文化的。
5. **re-enactment**：再次展現。
6. **dresses up**：化裝。

UNDERSTANDING THE TEXT

1 **Are the following sentences true (T) or false (F)? Correct the false ones.**

	T	F
a. Columbus Day is celebrated only in the United States.	☐	☐
b. Christopher Columbus wanted to reach the West by sailing to the East.	☐	☐
c. The King and Queen of Spain gave Columbus two ships: the *Santa Maria* and the *Pinta*.	☐	☐
d. Columbus was an Italian scientist.	☐	☐
e. He discovered the New World on 12 October, 1492.	☐	☐
f. Most Americans celebrate Columbus Day with patriotic speeches.	☐	☐
g. In San Francisco, California, there is a re-enactment of Columbus' discovery.	☐	☐

2 **Odd one out!**

Circle the word in each group that doesn't belong to the same group.

a. sea   round   lake   ocean   river

b. American   coast   British   Spanish   Italian

c. sailor   navigator   beach   explorer   teacher

d. city   town   horizon   village   metropolis

Now use the circled words to fill in the gaps.

a. Genoa is a city on the north-west ....

b. Columbus noticed that ships seemed to go under the ....

c. He was convinced that the world was ....

d. The sailors get into a boat and row to the ....

3 **Which of these men were great explorers? Circle their names.**

Marco Polo

King Arthur

Magellan

Robin Hood

Sir Francis Drake

Captain Cook

King Ferdinand

Columbus

Name a famous explorer from your country.

What did he discover or explore?

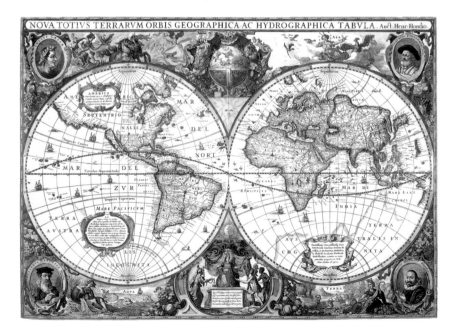

$\mathcal{H}$alloween [1]

$\mathcal{H}$alloween is celebrated on 31 October. It is an exciting event in the United States and in Great Britain. Every American calendar has Halloween marked on it.

Halloween has ancient Celtic origins. On the Celtic calendar Samhain was the last day of summer and the last day of the year, 31 October.

The Celtic priests, called Druids, practised[2] religious rituals[3] and magic on Samhain. They also predicted [4] the future.

On this day the Celts made big fires and dressed in scary[5] costumes[6]. They wanted to frighten the evil spirits. They dressed as ghosts, skeletons[7] and witches [8]. They believed that ghosts came out of their tombs on the night of 31 October.

Samhain also became a harvest festival after the Roman invasion[9] of

---

1. **Halloween**：萬聖節前夕。
2. **practised**：遵循。
3. **rituals**：儀式。
4. **predicted**：預言。
5. **scary**：駭人的。

6. **costumes**：服裝。
7. **skeletons**：骨架。
8. **witches**：
9. **invasion**：侵略。

An original way to use an enormous pumpkin!

British and American Festivities

Britain in 43 AD. Christian practices replaced pagan practices. The Christians called 1 November All Hallows' Day, the day of All Saints. The evening of 31 October was called All Hallows' Eve [1]. This became Halloween.

At Halloween American children in elementary schools take their costumes and masks to school. Typical [2] Halloween costumes are the witch, ghost, skeleton, monster, vampire [3] or alien [4]. Young people have fun making their own costumes, but some prefer to buy them. In the afternoon the children put on their costumes and have a Halloween party at school. Schools are decorated with [5] pumpkins [6], ghosts, witches and bats.

It is a popular tradition [7] to make jack o'lanterns [8] out of pumpkins. People put them in front of the windows of their homes. The jack o' lantern is of Celtic origin too.

There are funny Halloween games such as

A Halloween costume.

1. **Eve** : 前夜。
2. **typical** : 典型的。
3. **vampire** : 吸血鬼。
4. **alien** : 外星人。
5. **are decorated with** : 裝飾。
6. **pumpkins** :
7. **tradition** : 傳統。
8. **jack o'lanterns** :（把南瓜挖空雕成人面形的）傑克燈。

Halloween

'bobbing for apples [1]'. Favourite Halloween foods are candied apples [2], nuts, liquorice [3], popcorn [4] and pumpkin pie.

American teenagers have a Halloween party in the evening. The party is usually in the school gymnasium [5] and everyone wears a costume and mask. The best costume usually wins a prize.

Bobbing for apples.

Decorated jack o'lanterns.

1. **bobbing for apples**：咬蘋果遊戲。
2. **candied apples**：蜜餞蘋果。
3. **liquorice**：甘草。
4. **popcorn**：爆玉米花。
5. **gymnasium**：體育館。

British and American Festivities

In the 19th century Irish immigrants [1] took their Halloween customs to the United States. They introduced [2] the custom of 'trick-or-treating [3]'. American and British children and teenagers go 'trick-or-treating' on Halloween evening. They go from house to house in their costumes and ring doorbells [4]. When the door opens they shout, 'Trick or treat?' People usually give them sweets or money. If not, the children play a trick! They write on windows with soap or spray [5] shaving cream [6] on cars and people!

People decorate their homes with jack o'lanterns.

1. **immigrants**：移民。
2. **introduced**：引進，傳入。
3. **trick-or-treating**：不請吃就搗蛋 (遊戲)。
4. **ring doorbells**：按門鈴。
5. **spray**：噴灑。
6. **shaving cream**：

UNDERSTANDING THE TEXT

1 **Choose the correct answer.**

a. Halloween is celebrated on 31 October in
- [ ] Great Britain only
- [ ] the United States and Great Britain
- [ ] the United States only

b. The ancient origins of Halloween are
- [ ] Celtic
- [ ] Irish
- [ ] Roman

c. The word Halloween comes from
- [ ] a Roman word
- [ ] All Hallows' Eve
- [ ] an Irish song

d. American children in elementary schools
- [ ] have a Halloween party at school
- [ ] make Halloween cards
- [ ] sing Celtic songs

e. American teenagers
- [ ] don't celebrate Halloween
- [ ] don't go to school on 31 October
- [ ] have a party on Halloween evening

f. On Halloween evening American and British children and teenagers put on their costumes and masks and
- [ ] go 'trick-or-treating'
- [ ] dance in the streets
- [ ] write on doors with soap

The Past Simple（過去時）

The Past Simple of a verb is often used to tell a story. Use the verbs in the pumpkin to fill in the gaps in the sentences below. Use the Past Simple.

make   be
take
give   believe
come   dress
practise   become
wear

a. Samhain .... the last day of summer.

b. The Druids .... religious rituals and magic.

c. The Celts .... big fires and .... in scary costumes.

d. They .... that ghosts .... out of their tombs.

e. After 43 AD Samhain also .... a harvest festival.

f. Irish immigrants .... their Halloween customs to the United States.

g. The people .... the children sweets.

h. They .... frightening costumes.

3 **Match the name with the correct mask.**

witch   skeleton   devil

Batman   alien

a. ....

b. ....

c. ....

d. ....

e. ....

Which mask do you prefer? Why?

Will you wear a costume and mask for Halloween? If so, describe them.

Listen to the second and third paragraphs of page 16 and fill in the missing words.

At Halloween American .... in elementary schools take their costumes and masks to .... Typical Halloween costumes are the ...., ghost, skeleton, monster, vampire or alien. Young .... have fun making their own costumes, but some prefer to .... them. In the .... the children put on their costumes and have a Halloween .... at school. Schools are decorated .... pumpkins, ghosts, witches and bats.

It is a popular tradition to .... jack o'lanterns out of pumpkins. People put them in front of the .... of their homes. The jack o'lantern is of .... origin too.

American Pumpkin Pie

Invite your friends to a Halloween party. To make your party a big success, make an American pumpkin pie! It's delicious and easy to make! Here's a recipe [1] for four people.

Ingredients [2]

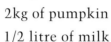

2kg of pumpkin

1/2 litre of milk

4 eggs

4 tablespoons of flour [3]

1 teaspoon of baking powder [4]

100g of butter

1/4 teaspoon of salt

1/4 teaspoon of cinnamon powder [5]

Utensils [6] to use

a big saucepan [7]

a tablespoon

a teaspoon

a wooden spoon

a big bowl

a non-stick pie dish

1. **recipe**：食譜。
2. **ingredients**：材料。
3. **flour**：麵粉。
4. **baking powder**：發酵粉。
5. **cinnamon powder**：肉桂粉。
6. **utensils**：用具。
7. **saucepan**：鍋。

Cut the pumpkin into small pieces and remove the peel [1]. Put the pieces into a big saucepan and add the milk. Cook the mixture for 30 minutes and don't forget to mix it while it's cooking.

Now break the eggs into a bowl. Add the flour, baking powder, butter, salt and cinnamon powder. Mix these ingredients well. Add the pumpkin and milk mixture to the ingredients in the bowl and mix. Put the mixture into a non-stick pie dish. Put the dish into the oven [2] (180°C) for one hour. Serve [3] the pumpkin pie cold. Happy Halloween!

Pumpkin cut into small pieces.

1. **peel**：瓜果皮。
2. **oven**：烤箱。
3. **serve**：端上。

Guy Fawkes' Night

The fifth of November is Guy Fawkes' Night or Bonfire [1] Night. This is a British festivity.

The story of Guy Fawkes goes back to the early 1600's. During this period there were religious problems between Protestants [2] and Catholics [3] in Britain. King James I was a Protestant and he passed severe laws against Catholics. They were not permitted to have religious services.

A group of 12 Catholics decided to kill King James I and destroy the Parliament Building! They planned to blow up [4] the Houses of Parliament on 5 November 1605, when the King was present. This was called the Gunpowder Plot.

The leader of the plot was Robert Catesby. The plotters [5] put 30 barrels of explosives [6] in the cellars [7] under the Parliament Building.

Guy Fawkes was an expert with explosives. His responsibility was to guard the barrels of explosives and light the fuse [8] on 5 November.

---

1. **bonfire** : 篝火。
2. **Protestants** : 新教徒。
3. **Catholics** : 天主教徒。
4. **blow up** : 炸毀。

5. **plotters** : 密謀者。
6. **explosives** : 炸藥。
7. **cellars** : 地窖。
8. **light the fuse** :

British and American Festivities

The King's soldiers discovered the plot! Guy Fawkes was immediately arrested and tortured [1]. The other plotters were found three days later. Guy Fawkes and the others were hanged [2].

On the night of 5 November 1605 many people in London were happy because the plot was discovered. To celebrate they started bonfires in the street. Someone made an effigy [3] of Guy Fawkes and burned it.

Ever since that day the British have celebrated Guy Fawkes' Night. Today young people in Britain make a Guy with old clothes and fill him with newspaper. Then they go around the streets with the Guy and ask for 'a penny for the Guy?' With this money they buy fireworks [4].

On the night of 5 November there are fireworks and big bonfires to burn the Guy. Some people have bonfires in their back gardens. Young

1. **tortured** : 折磨。

2. **hanged** :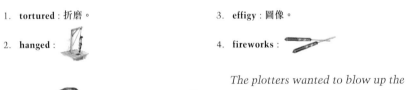

3. **effigy** : 圖像。

4. **fireworks** :

The plotters wanted to blow up the Houses of Parliament on 5 November 1605.

'A penny for the Guy?'

people love the noise, excitement and colourful fireworks.

On this night they eat toffee apples [1].

In Lewes, Sussex, there is a big public festival on Guy Fawkes' Night. People dress in historic [2] costumes. The Guy is burned on top of a hill on an enormous bonfire. There are also brilliant [3] fireworks.

Here is a rhyme [4] about Guy Fawkes:

> *Remember, remember*
> *The fifth of November*
> *Gunpowder, treason [5] and plot.*
> *I see no reason*
> *why gunpowder and treason*
> *Should ever be forgot.*

1. **toffee apples**：塗太妃糖的蘋果。
2. **historic**：古代的。
3. **brilliant**：燦爛的。
4. **rhyme**：韻文。
5. **treason**：叛國。

UNDERSTANDING THE TEXT

 1 Choose the correct answers.

a. Guy Fawkes' Night is *an American / a British* festivity.

b. King James I was a *Protestant / Catholic* and he passed severe laws against *his people / Catholics*.

c. A group of 12 *soldiers / Catholics* planned to *enter / blow up* the Houses of Parliament on 5 November 1605.

d. Guy Fawkes' responsibility was to *make / guard* the barrels of explosives.

e. When the *King's soldiers / King* discovered the plot, Guy Fawkes and the other plotters *were hanged / escaped*.

f. Today young people in *Britain / America* make a *Guy / King* with old clothes.

g. On the *night / day* of 5 November there are fireworks and big bonfires to burn the Guy.

2 Fill in the gaps with the words below. Some words can be used twice.

Who	Why	When	Where	What

a. .... was King of Britain in 1605?
King James I was King in 1605.

b. .... did the plotters plan to blow up the Parliament Building?
They planned to blow it up on 5 November 1605.

c. .... was the leader of the plot?
Robert Catesby was the leader.

d. .... did the plotters put the barrels of explosives?
They put them in the cellars under the Parliament Building.

e. .... were Guy Fawkes and the other plotters hanged?

They were hanged because they wanted to blow up the Parliament Building and kill King James I.

f. .... do some people have bonfires?

They have them in their back gardens.

g. .... happens in Lewes, Sussex on Guy Fawkes' Night?

There is a public festival with an enormous bonfire and fireworks.

3 **Make a list of things you need to make a Guy. Look at the photograph on page 27. The words in the bonfire will help you.**

hat
newspaper
shirt   glasses
paper bag
trousers   jumper
hair
dress   shoes

4 **Go back to page 27 and read the rhyme about Guy Fawkes. What do you think its message is?**

CHAPTER FOUR

Thanksgiving

Thanksgiving is a very important American festivity and it is celebrated on the last Thursday of November. However, schools and many shops and offices are closed for four days: Thursday, Friday, Saturday and Sunday.

Thanksgiving is a special day for families to be together and to thank God for all they have. Americans travel great distances to be with their families on this occasion [1].

Why is Thanksgiving such an important day? The tradition started with the Pilgrims [2], the founders [3] of America. The Pilgrims left Great Britain in 1620 because of religious persecution [4]. They wanted to start a new life in America and practise their religion in freedom [5].

One hundred men, women and children left Britain on a small ship called the *Mayflower*. Their sea voyage was very difficult. Many Pilgrims died during the voyage.

They arrived on the north-east coast of North America in December 1620 and founded[6] Plymouth. The area was a wilderness. It was almost

1. **occasion** : 時刻。
2. **pilgrims** : 朝聖者。
3. **founders** : 建立者。

4. **persecution** : 迫害。
5. **freedom** : 自由。
6. **founded** : 創建。

winter and they had no homes and little food. They immediately built small homes, but it was too late to cultivate[1] crops[2].

The winter was very cold and harsh[3]. Almost half of the Pilgrims died because their living conditions were very bad.

The friendly Wampanoag Indians helped them during the long winter.

In the spring the Pilgrims met an Indian called Squanto. He explained how to grow corn, hunt and live in the wilderness.

Soon the Pilgrims and the Indians became good friends.

The Pilgrims worked hard and cultivated crops. The summer harvest was excellent. By November

1. **cultivate** : 種植。

2. **crops** :

3. **harsh** : 嚴酷的。

British and American Festivities

1621 everyone had food and a home. There was hope for the future.

Governor William Bradford, the Pilgrim leader, decided to celebrate with a dinner for the Pilgrims and the Indians. He wanted to give thanks to God. This was the first Thanksgiving dinner and it lasted [1] for three days!

Today the traditional Thanksgiving meal is similar to the first.

Roast turkey is part of the traditional Thanksgiving dinner.

1. **lasted** : 持續。

Thanksgiving

People eat roast turkey, sweet potatoes, corn, cranberry sauce [1] and pumpkin pie. Most families start the meal with a prayer [2].

The long Thanksgiving weekend is the perfect time to start Christmas shopping! Big stores and shops are open on Friday, Saturday and Sunday for the Christmas shoppers.

A float in a Thanksgiving parade in New York.

1. **cranberry sauce** : 越橘醬。
2. **prayer** : 祈禱。

UNDERSTANDING THE TEXT

1 **Are the following sentences true (T) or false (F)? Correct the false ones.**

	T	F
a. Thanksgiving is celebrated on the last Thursday of November.	☐	☐
b. Americans travel great distances to be with their families on Thanksgiving Day.	☐	☐
c. The tradition started in Great Britain.	☐	☐
d. The Pilgrims were the founders of America.	☐	☐
e. Their first winter was very cold but they had lots of food.	☐	☐
f. Squanto helped the Pilgrims to build the first houses.	☐	☐
g. The first Thanksgiving dinner was in November 1621.	☐	☐
h. During the long Thanksgiving weekend, many people start Christmas shopping.	☐	☐

2 **Unscramble the anagrams**（變形詞） **and then match them with their meanings.**

a. REAFMOYWL　　................................. ☐

b. LESWRIDENS　　................................. ☐

c. MSIPLGIR　　................................. ☐

d. ARHSH　　................................. ☐

e. UNSTOQA　　................................. ☐

f. OSRPC　　................................. ☐

g. OLHPTUYM　　................................. ☐

1. Land that is not cultivated, where wild animals live.

2. Very unpleasant.

3. The Pilgrims' ship.

4. Fruit, vegetables and corn.

5. The founders of America.

6. A friendly Indian.

7. The colony founded by the Pilgrims.

3 **You are a young pilgrim in Plymouth. It is the day after Thanksgiving and you want to write a letter to your grandmother in Britain. Write full sentences with the notes given. Use the Past Simple tense** （過去時）.

Dear Grandmother,

America / arrive / in / I / one / ago / year.

be / The / very / winter / cold / little

and / be / there / food.

people / Many / die.

the / spring / kind / Indians / In / help / some / us.

our / They / friends / become / good.

build / small / We / houses / crops / and / cultivate.

The / good / be / very / this / harvest / year.

invite / Indians / Yesterday / we / the / Thanksgiving / a / to / dinner.

for / God / everything / thank / We.

                    Love from,

                    ....

Pilgrim's Apple Crumble [1]

Do you want to make an original [2] dessert [3] for Thanksgiving? Try the Pilgrim's Apple Crumble, a recipe from long ago! It's easy to make and delicious to eat.

Ingredients
250g of flour
150g of brown sugar
100g of butter (at room temperature)

To make the fruit mixture you need:
900g of sliced [4] apples
2 tablespoons of brown sugar
1 cinnamon stick
1/4 cup of water

Utensils to use
a big bowl
a big saucepan
a wooden spoon
a tablespoon
a teaspoon
a non-stick pie dish

Make the crumble first. Put the flour into a bowl. Add the butter and mix with your hands. When the mixture is crumbly [5], add the sugar and mix well.

1. **apple crumble** : 蘋果點心。
2. **original** : 新穎的。
3. **dessert** : 甜食。
4. **sliced** :
5. **crumbly** : 易碎的。

Then make the fruit filling [1]. Put the sliced apples, sugar and cinnamon stick into a big saucepan. Add the water and cook until the apples are soft.

When the fruit is ready put it into a non-stick pie dish and cover the fruit with the crumble. Put the pie dish into the oven (200°C) and bake [2] the crumble for 30-40 minutes. Remember, the top must be light brown. Happy Thanksgiving!

1. **filling** : 餡。   2. **bake** : 烘烤。

CHAPTER FIVE

Christmas

On 25 December Great Britain and the United States (and many other countries) celebrate Christmas. The word Christmas comes from the Old English 'Cristes maesse', Christ's mass. On this day Christians celebrate the birth of Christ.

Kindness and giving are the spirit of Christmas. Charles Dickens, the famous British writer, wrote *A Christmas Carol*[1], a story about the true meaning of Christmas.

Why is Christmas celebrated on 25 December?

Christmas comes from two pagan festivals. People celebrated the winter solstice[2] on 21 or 22 December. The Roman emperors chose 25 December as the birthday of the sun (natalis solis).

After the winter solstice, the days become longer and the sun is higher in the sky. In ancient Rome this was the season of Saturnalia. Saturnalia was a time of merrymaking[3]. During Saturnalia the Romans exchanged[4] presents.

---

1. **carol** : 聖誕聖歌。
2. **winter solstice** : 冬至。
3. **merrymaking** : 尋歡作樂。
4. **exchanged** : 交換。

Christmas

The pagan tribes of northern Europe enjoyed a 12-day winter festival called Yule. Yule had its own traditions of Yule cakes, fir trees [1], holly [2], mistletoe [3] and presents. Fir trees and holly are still a symbol [4] of Christmas. Today a branch of mistletoe in the house has another meaning: when a boy and girl meet under the mistletoe they usually kiss! The custom of kissing under the mistletoe is an ancient Celtic rite [5].

Medieval [6] Christmas was a long event. It lasted twelve days like the Yule festival. Celebrations started on 25 December and ended on the night of 6 January.

Christmas is the biggest holiday on the American and British calendar. However, in Scotland, New Year is more important. In Great Britain and the United States people prepare for Christmas weeks before 25 December.

Cities and towns are beautifully decorated with Christmas symbols: the Christmas tree, Santa Claus, colourful lights and much more. Shop windows are full of

1. **fir trees :**

2. **holly :**

3. **mistletoe :**
4. **symbol** : 象徵。
5. **rite** : 儀式。
6. **Medieval** : 中世紀的。

British and American Festivities

presents for everyone. People are busy buying presents. There are Santa Clauses everywhere. On street corners people sing Christmas carols. In Britain many children go carol singing with an adult. They go from house to house and sing. They usually receive money for their singing. They use this money to buy presents. There is a joyous atmosphere[1].

Christmas today respects many of the old traditions and has added new ones.

The modern Christmas tree originated in[2] western Germany long ago. The Germans put up a fir tree in their homes and decorated it with biscuits[3] and candles[4]. German settlers took this tradition to North America in the 17th century. By the 19th century Christmas trees were very popular in the United States.

The first modern Christmas tree came from Germany.

1. **joyous atmosphere**：喜慶的氣氛。
2. **originated in**：起源於。
3. **biscuits**：餅乾。
4. **candles**：蠟燭。

40

Christmas

In Britain the Christmas tree became popular after 1840. The German Prince Albert, Queen Victoria's husband, took the Christmas tree to the British Royal Family. Soon the Christmas tree became popular all over Great Britain. There is a big tree every year in Trafalgar Square, London.

Today America's most famous Christmas tree is in the Rockefeller Center, New York City.

The Empire State Building in New York City is one of the tallest buildings in the world. At Christmas the top of the building is illuminated[1] with red and green lights!

Today almost every Christian family has a decorated Christmas tree at home or in the garden. Some families put up the tree on Christmas

America's most famous Christmas tree.

1. **illuminated**：照明。

British and American Festivities

Eve and others put it up at the beginning of December. Many families put a Nativity [1] scene [2] under the Christmas tree. St. Francis of Assisi created the first Nativity scene in the 1200's!

Many Americans wear 'Christmas clothes' during the Christmas season. These are clothes with Christmas colours and symbols.

Christmas crackers [3] are an old tradition. They were invented in London. Two people pull the cracker

A Nativity scene.

until it 'bangs [4]' and opens. Inside there is usually a small present, a paper hat and a joke.

In the United States there are 'Christmas shops' in many cities. They sell all types of Christmas trees, ornaments [5] and other decorations [6] for the home. 'Christmas shops' are open all year long, so you can buy your Christmas tree in June!

It is usual to send Christmas cards to relatives [7] and friends. The first Christmas card was designed in

1. **Nativity**：耶穌的出生。
2. **scene**：景象。
3. **crackers**：爆竹。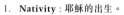
4. **bangs**：發出巨響。

5. **ornaments**：裝飾物。
6. **decorations**：裝飾物。
7. **relatives**：親戚。

42

Christmas

Britain by John Callcott Horsley in 1843. A thousand copies of the card were printed in London. The design was a family party with the words 'A Merry Christmas and a Happy New Year to you.'

In the mid-19th century a shop owner in Albany, New York, made a card that said, 'Christmas greetings from Pease's Great Variety Store.' This was the beginning of the big Christmas card industry [1].

A Christmas card.

The figure [2] of Santa Claus comes from St. Nicholas, a 4th century Christian bishop of Asia Minor. He was famous for his generosity and kindness. The Dutch settlers of New York brought the tradition of Sinterklaas (Santa Claus) to America.

Father Christmas, or Santa Claus, is an important part of the Christmas tradition. Children write letters to him and ask for special presents.

Today we see Santa Claus as an old man with a white beard, dressed in a red suit. He drives a sleigh [3] pulled by reindeer [4] and comes down the chimney [5] with toys for the children. This image of Santa Claus was created by the American cartoonist [6] Thomas Nast in 1863.

American and British children put out a stocking [7] on Christmas Eve because they hope to receive presents from Santa Claus. Then they go to

1. **industry**：行業。
2. **figure**：形象。
3. **sleigh**：雪橇。
4. **reindeer**：
5. **chimney**：
6. **cartoonist**：漫畫家。
7. **stocking**：長筒襪。

British and American Festivities

bed early because they want Santa Claus to come. Many Christians go to church services at midnight and others go on Christmas morning. Christmas presents are usually opened on Christmas morning. After Christmas dinner many families listen to the traditional Queen's Speech in Britain and watch special Christmas television programmes. Children play with their new toys.

In Britain some people go for a Christmas Day swim in the sea or in a lake. In Hyde Park in London some courageous people swim in the Serpentine!

Christmas stockings.

The First Noël

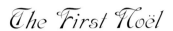

The first Noël the angel did say
was to certain poor shepherds in
　fields as they lay;
In fields where they lay, keeping
　their sheep,
On a cold winter's night that was so
　deep:

Noël, Noël, Noël, Noël,
Born is the King of Israel!

They looked up and saw a star,
Shining in the east, beyond them far;
And to the earth it gave great light,
And so it continued both day and
　night:

And by the light of that same star,
Three Wise Men came from country
　far;
To seek for a king was their intent [1],
And to follow the star wherever it
　went:

Traditional.

1. **intent** : 目的。

O Come, All Ye Faithful [1]

O come, all ye faithful,
Joyful and triumphant [2]
O come ye, O come ye to
    Bethlehem;
Come and behold him
Born the King of Angels:

O come, let us adore [3] him,
O come, let us adore him,
O come, let us adore him,
    Christ the Lord!

God of God
Light of light,
Lo! he abhors [4] not the Virgin's
    womb [5];
Very God,
Begotten not created:

Sing, choirs [6] of angels,
Sing in exultation [7],
Sing, all ye citizens of heaven
    above;
Glory to God
In the highest:

Traditional.

1. **faithful** : 忠誠的。
2. **triumphant** : 得勝的。
3. **adore** : 崇拜。
4. **abhors** : 厭惡。
5. **womb** : 子宮。
6. **choirs** : 詩班。
7. **exultation** : 狂喜。

UNDERSTANDING THE TEXT

1 Choose the correct answer.

a. Christmas comes from
- [ ] a Celtic tradition
- [ ] two pagan festivals
- [ ] Germany

b. The Roman emperors chose 25 December as
- [ ] the birthday of the sun
- [ ] the birthday of the emperor
- [ ] the winter solstice

c. The pagan tribes of northern Europe enjoyed a 12-day winter festival called
- [ ] Saturnalia
- [ ] Medieval Christmas
- [ ] Yule

d. The modern Christmas tree originated in
- [ ] Great Britain
- [ ] Germany
- [ ] New York

e. The first Christmas card was
- [ ] designed in Britain in 1843
- [ ] designed in America in 1863
- [ ] printed in Albany, New York

f. The figure of Santa Claus comes from
- [ ] the Yule festival
- [ ] St. Francis
- [ ] St. Nicholas

g. The modern image of Santa Claus was created by
- [ ] an American cartoonist
- [ ] Dutch settlers
- [ ] a shop owner in Albany, New York

Christmas Traditions

Food

The traditional Christmas dinner consists of [1] roast turkey [2] or roast ham [3], vegetables, salad and special desserts. In Britain Christmas cake, Christmas pudding and mince pies [4] are the typical desserts. Fruit cake, pumpkin pie and special Christmas cookies [5] are favourite American desserts.

 Can you name the traditional Christmas foods on page 49? The words in the tree will help you.

fruit cake

roast ham

roast turkey

Christmas pudding

mince pies

Christmas cookies

---

1. **consists of** : 包括。
2. **roast turkey** : 烤火雞。
3. **roast ham** : 烤火腿。

4. **mince pies** : 肉餡餅。
5. **cookies** : (美國) 餅乾。

a. ....

b. ....

c. ....

d. ....

e. ....

f. ....

Trees

The traditional Christmas tree is an evergreen[1], because green is the colour of life. Most people have green Christmas trees but others decorate the brown branches of a tree with ornaments and create a very original tree. Look at these unusual Christmas trees.

a. A child's tree.

c. A dog's tree.

b. A tree made of twigs[2].

1. **evergreen** : 常青樹。

2. **twigs** : 細樹枝。

d. A decorated tree trunk[1].

e. Snow-covered fir trees.

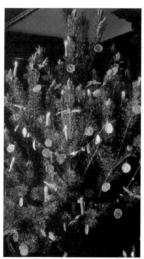

f. A tree with
red ribbons.

h. A Christmas
cookie tree.

g. A tree with
orange slices[2].

1. **trunk**：樹幹。　　　　2. **slices**：薄片。

1 **Which Christmas tree(s):**

a. has ornaments that you can eat? ....

b. has teddy bears on it? ....

c. belongs to a dog? ....

d. is decorated with ribbons? ....

e. isn't green? ....

f. do you like best? ....

Why? ....

....

2 **Describe your Christmas tree.**

....

....

....

....

....

Christmas around the World

C hristmas is celebrated in different ways around the world. People in Mexico and Australia can celebrate Christmas on the beach because it's summer in December! Most people in northern Europe enjoy a 'white Christmas' because there's a lot of snow. There are many Christmas markets in December, especially in central and northern Europe. Outdoor Christmas markets sell all types of Christmas decorations, ornaments for Christmas trees and presents.

1 Look at these Christmas pictures from different countries and match the pictures with the correct countries.

| Germany | Lapland [1] | U.S.A. | Italy |
| Great Britain | Mexico | | |

a. ....

b. ....

c. ....

1. **Lapland**：(北歐)拉普蘭。

d. ....

e. ....

f. ....

2 Look at the shop window below. Label the things you can see. The words in the Santa Claus will help you.

ornament
rocking horse
Santa Claus
teddy bear
Nativity scene
Christmas tree

a. ....

b. ....

c. ....

d. ....

e. ....

f. ....

New Year's Eve

On 31 December everyone celebrates the end of the old year and the beginning of the new.

In Scotland New Year's Eve is called Hogmanay. It is the most important celebration of the year. In Edinburgh there is an immense [1] street celebration on New Year's Eve.

In America and Britain many people like going to parties or organising them in their homes. Some people prefer to celebrate in restaurants or night clubs. Young people celebrate at home or at a disco. Others go to masked balls [2] in costumes and masks. Everyone takes off their mask at midnight.

Cheers [3], noise, music, dancing, colourful decorations, festive food and drink are all part of New Year's Eve. People like throwing confetti [4] on the last night of the year.

1. **immense** : 巨大的。
2. **balls** : 舞會。
3. **cheers** : 喝彩。
4. **confetti** : 五彩紙屑。

British and American Festivities

People wear their best party clothes on this exciting night. At parties everyone wears funny paper hats and blows toy horns [1]. Parties usually begin after 9 pm and continue until the next morning. At midnight everyone joins hands and sings the old Scottish song 'Auld Lang Syne.'

In New York City a favourite place to go on New Year's Eve is Times Square. At midnight the words 'Happy New Year' appear on an electronic sign [2]. Bells and sirens [3] ring, people cheer and there is a lot of noise!

People wait for Big Ben to strike midnight.

In London many people go to celebrate in Trafalgar Square. Everyone in Britain waits for Big Ben to strike midnight. Then there is a lot of noise. People sing 'Auld Lang Syne', kiss each other and cheer.

When we make a lot of noise on New Year's Eve we are following ancient traditions. Ancient civilizations [4] made noise to frighten evil spirits [5] of the past year.

On New Year's Eve the Americans and the British

1. **toy horns :**
2. **electronic sign :** 電子標誌。
3. **sirens :** 汽笛。
4. **civilizations :** 文明。
5. **evil spirits :** 邪惡的精靈。

make New Year's resolutions [1] or promises. They promise to get rid of [2] bad habits during the new year. Some typical children's resolutions are: 'I resolve to do my homework' or 'I resolve to clean my room every day'. Children write down their New Year's resolutions, sign their name and give the paper to their parents.

Some typical adult's resolutions are: 'I resolve to stop smoking' or 'I resolve to go on a diet [3]'. Most people don't keep their promises [4], but a few do! In some American cities office workers throw their old calendars out of the office windows on 31 December. They are throwing the old year away! It is incredible [5] to see so much paper flying about. By 5:30 pm the streets are covered with office calendars.

At the end of the year astrologists [6] predict the future for all the signs of the zodiac [7]. Some astrologists predict catastrophic [8] events!

1. **resolutions** : 決心。
2. **get rid of** : 除掉。
3. **go on a diet** : 節省。
4. **keep their promises** : 遵守承諾。
5. **incredible** : 不可思議的。
6. **astrologists** : 占星學家。
7. **zodiac** : 黃道帶。
8. **catastrophic** : 災難的。

UNDERSTANDING THE TEXT

 Fill in the gaps with the correct words from the calendar.

December

office   parties

Hogmanay

calendars

favourite   Big Ben

balls

Times Square

promises

important

a. In Scotland New Year's Eve is called .... and it is the most .... celebration of the year.

b. On New Year's Eve people like going to ...., or to masked ....

c. In New York City a .... place to go is ....

d. In London everyone waits for .... to strike midnight.

e. New Year's resolutions are .... to get rid of bad habits during the new year.

f. In some American cities .... workers throw their old .... out of the office windows.

Find the names of three cities, two squares and a famous clock mentioned in this chapter.

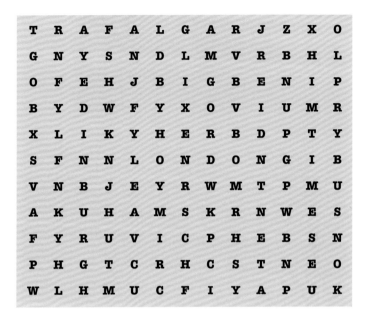

T	R	A	F	A	L	G	A	R	J	Z	X	O
G	N	Y	S	N	D	L	M	V	R	B	H	L
O	F	E	H	J	B	I	G	B	E	N	I	P
B	Y	D	W	F	Y	X	O	V	I	U	M	R
X	L	I	K	Y	H	E	R	B	D	P	T	Y
S	F	N	N	L	O	N	D	O	N	G	I	B
V	N	B	J	E	Y	R	W	M	T	P	M	U
A	K	U	H	A	M	S	K	R	N	W	E	S
F	Y	R	U	V	I	C	P	H	E	B	S	N
P	H	G	T	C	R	H	C	S	T	N	E	O
W	L	H	M	U	C	F	I	Y	A	P	U	K

 Listen to the dictation carefully. Then listen to it again and fill in the gaps with the correct words.

People like .... to parties on New Year's Eve. They wear their best .... clothes. There is a lot of noise everywhere. People ...., sing, eat and drink. At midnight, bells .... and people cheer. In Edinburgh, Scotland, there is a big .... celebration.

At the .... of the year astrologists predict the future and some people .... New Year's resolutions.

New Year's Day

'Happy New Year!'

'The same to you and many more.' This familiar[1] greeting is heard throughout Britain and the United States on 1 January.

People have always celebrated the New Year. In ancient civilizations the calendar was based on the seasons. The Egyptian New Year began when the Nile River overflowed[2]. For the early Britons and the Romans the New Year began on the first day of spring.

In 45 BC the Roman Emperor Julius Caesar created a calendar with 1 January as the first day of the new year. This calendar is still used today. January comes from 'Janus,' the Roman god of beginnings and endings.

New Year's traditions come from different cultures. Many people give presents and send cards on New Year's Day. This was part of a Roman and old English tradition. Shops and offices give calendars and small presents to their clients.

---

1. **familiar**：熟悉的。
2. **overflowed**：泛濫。

New Year's Day

In Britain most families have a big lunch and spend a quiet day at home.

One of the first American football teams.

In the United States many families have 'Open House' on New Year's Day. It is a custom introduced by George Washington, the first President of the United States. During 'Open House' the front door of your home is open all day long. Friends and relatives come to say 'Happy New Year!' They eat and drink something and then leave. Many clubs and organisations have 'Open House' too.

On New Year's Day most of the United States is covered with snow. However, in California and in the southern states it is warm and sunny. In these sunny places there are parades and football games. These football games are called Bowl Games. Each region has its parade and football game:

A modern American football team.

FOOTBALL GAME	PLACE
Rose Bowl	*Pasadena, California*
Orange Bowl	*Miami, Florida*
Cotton Bowl	*Dallas, Texas*
Sugar Bowl	*New Orleans, Louisiana*

British and American Festivities

The Pasadena 'Tournament [1] of Roses' parade is the biggest and oldest New Year's Day event. In Pasadena more than three million people go to watch the parade! More than 70 million Americans watch it on television.

A float made of fresh flowers.

Every year there are about 60 spectacular [2] floats made of fresh flowers. The floats show favourite storybook characters [3] and animals. The queen of the parade is called the Citrus [4] Queen, because so many citrus fruits grow in southern California.

It takes about a year to organise the Pasadena 'Tournament of Roses' parade! The entire city participates in this extraordinary [5] event.

When the parade ends everyone goes to the football stadium [6] to watch the Rose Bowl game, the biggest sports event of the year.

Another famous New Year's Day parade is the Macy's Day Parade in New York City. Many Americans watch this parade on television in the morning and in the afternoon they watch a football game.

The 'Tournament of Roses' parade.

---

1. **Tournament**：錦標賽。
2. **spectacular**：壯觀的。
3. **characters**：人物。
4. **Citrus**：柑橘。
5. **extraordinary**：不尋常的。
6. **stadium**：體育場。

UNDERSTANDING THE TEXT

1 **Are the following sentences true (T) or false (F)? Correct the false ones.**

		T	F
a.	People started celebrating the New Year in 45 BC.	☐	☐
b.	The Roman Emperor Julius Caesar created a calendar with 1 January as the first day of the new year.	☐	☐
c.	In Britain many families have 'Open House' on New Year's Day.	☐	☐
d.	In California and in the southern states it is warm and sunny in winter.	☐	☐
e.	The Pasadena 'Tournament of Roses' is an important football game.	☐	☐
f.	The Rose Bowl game is the biggest sports event of the year.	☐	☐

2 **Look at this sentence:**

The Pasadena 'Tournament of Roses' parade is the *biggest* and *oldest* New Year's Day event.

Biggest **and** ***oldest*** **are superlatives**（最高級）**.**

Most superlatives are formed by adding -est to the adjective（形容詞）**:**

ADJECTIVE	SUPERLATIVE
young	youngest
small	smallest
rich	richest
cold	coldest
easy	easiest
noisy	noisiest
sunny	sunniest

Remember this grammar rule:

When the adjective ends in consonant（輔音）***y* for example *easy* change the *y* to *i* and add *-est*.**

Some superlatives are irregular（不規則的）**:**

ADJECTIVE	SUPERLATIVE
bad	worst
good	best

Fill in the gaps with the correct superlative from the two boxes.

a. New Year's Eve is noisy. It's the .... night of the year.

b. California is a sunny place. It's the .... place in America.

c. There were many young children in the parade, but John was the ....

d. A lot of Romans were rich, but Emperor Julius Caesar was the ....

e. Winter is the .... season of the year.

f. The biggest float won the prize because it was the ....

3 **Can you match the sport with its correct name?**

a. football   d. basketball

b. skiing   e. American football

c. volleyball   f. swimming

2 ☐

1 ☐

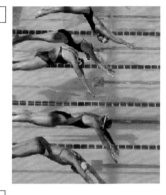

4 ☐

3 ☐

5 ☐

6 ☐

CHAPTER EIGHT

Chinese New Year

Chinese New Year is the most important festivity for the Chinese people in America and Great Britain. It is also called the Spring Festival [1] or the Lunar New Year [2].

The exact date on the Western calendar changes from year to year. However, Chinese New Year takes place between 1 January and 19 February. On the Chinese lunar calendar every month begins with the new moon.

Every year has an animal's name. These animals are the rat, ox, tiger, rabbit, dragon [3], snake, horse, ram [4], monkey, rooster [5], dog and pig (see page 69). A Chinese legend says that these twelve animals had a race. The first year was named after the rat, the winner. The other eleven years were named after the order [6] in which the animals arrived in the race. The clever rat jumped onto the ox's back then at the end jumped over the ox's head to arrive first!

The Chinese believe that a person born in a

1. **the Spring Festival** : 春節。
2. **the Lunar New Year** : 農曆新年。
3. **dragon** : 龍。

4. **ram** : 羊。
5. **rooster** : 雞。
6. **order** : 順序。

particular year has some of the characteristics [1] of that animal.

Celebrations in Chinese families last for about two weeks. The celebrations begin with traditional house cleaning. The Chinese get rid of old and useless things. They do this to throw away the misfortunes [2] of the past year.

A Chinese New Year card.

On Chinese New Year's Eve all family members enjoy a big, delicious meal. It is very important for the Chinese to be with their families on this occasion. Fish is always part of the dinner because it represents [3] abundance [4].

On New Year's Day all Chinese children wear new clothes with bright colours. Red is considered a lucky colour. Parents and relatives give children the traditional New Year's gift called 'Lai see' (lucky money). This money is put into bright red and gold envelopes. Red is a traditional colour for festivals, celebrations, weddings and birthdays.

In Britain a parade takes place in Soho, London. This is one of the biggest parades in an English-speaking country. Dragon or lion dancers [5] often lead the parade. In Chinatown, Soho there are many Chinese restaurants. It is possible to stop and eat typical Chinese food.

Chinese New Year was celebrated on 21 February, 1851 for the first time in San Francisco! This was during the California Gold Rush [6]. A lot of Chinese immigrants worked in California during the Gold Rush.

In big American cities such as San Francisco, New York, Honolulu and Houston, Chinese New Year is a major event with wonderful parades.

San Francisco, California, has the biggest Oriental [7] community outside

1. **characteristics**：特點。
2. **misfortune**：不幸。
3. **represents**：代表。
4. **abundance**：富裕。
5. **lion dancers**：舞獅人。
6. **Gold Rush**：淘金熱。
7. **Oriental**：東方的。

British and American Festivities

of Asia. This area of San Francisco is called Chinatown. There are many Chinese shops, restaurants and libraries in Chinatown. All street and shop signs are written in Chinese! During the Chinese New Year, Chinatown is decorated with beautiful ornaments.

Most of the costumes and masks come from Hong Kong. Every year there is a Miss Chinatown USA beauty and talent contest [1].

The lion dancers are always part of the festivities. The lion has a big head and long body made of cloth. The lion dance is accompanied by [2] drums, cymbals [3] and noisy firecrackers [4]. According to ancient traditions the great noise frightens away evil spirits.

The dragon is the most important figure of the Chinese New Year festivities and parades. The dragon is considered a lucky figure. A parade dragon can be 20 to 30 metres long! Sixty or more men move under a long cloth that represents the dragon's tail.

During the parade children represent the animals of the Chinese calendar. There are also acrobats [5] and musicians in beautiful costumes.

Every year the San Francisco parade attracts thousands of spectators [6]. It is a magnificent [7] event.

A ceramic Chinese dragon.

1. **contest** : 競賽。
2. **is accompanied by** : 由⋯伴奏。
3. **cymbals** :
4. **firerackers** :
5. **acrobats** : 雜技演員。
6. **spectators** : 觀看者。
7. **magnificent** : 壯麗的。

UNDERSTANDING THE TEXT

 Choose the correct answer.

a. Chinese New Year takes place
- [ ] between 1 January and 1 February
- [ ] on 1 January
- [ ] between 1 January and 19 February

b. On the Chinese lunar calendar every year
- [ ] has an animal's name
- [ ] ends with the new moon
- [ ] has a bird's name

c. On Chinese New Year's Day all Chinese children
- [ ] receive many sweets
- [ ] wear new clothes with bright colours
- [ ] play special games

d. Red and gold envelopes are used to give 'Lai see' which is
- [ ] a Chinese calendar
- [ ] a Chinese New Year card
- [ ] lucky money

e. The biggest Oriental community outside of Asia is in
- [ ] Chinatown in San Francisco
- [ ] Chinatown in New York
- [ ] Chinatown in London

f. The most important figure of the Chinese New Year festivities and parades is the
- [ ] monkey
- [ ] lion
- [ ] dragon

2 In this word square find the names of five things mentioned in this chapter.

A	C	E	G	I	K	M	O	Q	S	U
W	C	Y	B	D	F	R	A	T	H	J
L	A	H	N	P	R	O	T	V	X	Z
J	L	P	I	Z	U	T	M	B	O	P
N	E	R	S	N	Q	B	O	S	T	U
Q	N	G	D	R	A	G	O	N	H	L
O	D	N	A	Z	B	T	N	Y	C	X
D	A	W	E	V	F	U	O	G	T	H
S	R	I	R	J	Q	K	P	W	L	P
M	O	P	Z	C	T	L	O	B	N	D
T	R	C	L	D	G	N	H	S	P	O

3 Look at the Chinese Calendar on page 73. Find out which animal year you were born in!

YOUR DATE OF BIRTH

day	month	year

YOUR CHINESE SIGN

Do you like your sign? Why or why not?
What is your lucky colour?

The Chinese Calendar

Here are the twelve animals of the Chinese calendar and their personalities[1]:

1. The rat is happy, charming[2] and easily bored.
2. The ox is patient[3], shy and works a lot.
3. The tiger has a strong personality and likes adventure.
4. The rabbit is friendly, peaceful and shy.
5. The dragon is dynamic[4], loves freedom and is generous.
6. The snake is very wise and has a good sense of humour[5].
7. The horse works a lot and is ambitious[6].
8. The ram is gentle and loves beautiful things.
9. The monkey is clever, charming and creative[7].
10. The rooster loves friends and family and is punctual[8].
11. The dog is loyal, courageous and loves justice.
12. The pig is peaceful, strong and likes a quiet life.

1. **personalities** : 個性。
2. **charming** : 迷人的。
3. **patient** : 耐心的。
4. **dynamic** : 有活力的。
5. **humour** : 幽默。
6. **ambitious** : 有雄心的。
7. **creative** : 有創造力的。
8. **punctual** : 準時的。

CHAPTER NINE

Valentine's Day [1]

Valentine's Day, 14 February, is dedicated to people in love! On 14 February sweethearts [2] celebrate Valentine's Day with cards and presents. Cards can be funny, romantic or poetic [3]. Presents can be a box of chocolates, flowers or jewellery [4].

When and where did this festivity start?

The custom of celebrating Valentine's Day probably began with the Roman festival of Lupercalia on 15 February. During Lupercalia there were games and dancing. Every young man took the name of a young lady from an urn [5]. The lady was the young man's sweetheart for one year.

With the beginning of Christianity pagan customs were not permitted, but the festival continued. In the 7th century this festival was called St. Valentine's Day.

The origin of the name is still a mystery. Some historians say that the festival took its name from a Christian martyr [6] named Valentine. He died on 14 February in the year 270.

---

1. **Valentine's Day** : 情人節。
2. **sweethearts** : 戀人。
3. **poetic** : 充滿詩意的。
4. **jewellery** :
5. **urn** : 甕。
6. **martyr** : 殉道者。

Valentine's Day

Others say Valentine was put in prison by Emperor Claudius because he secretly married young couples. The Emperor did not like these secret marriages.

Some say that Valentine comes from the French word 'galantin' (a gallant or beau [1]).

A legend says that birds begin to mate [2] on 14 February!

The Roman conquerors [3] brought the celebration to England.

In England the pagan and Christian customs combined [4] to form the Valentine festivity.

The Valentine tradition was popular in Shakespeare's time.

Shakespeare's characters Romeo and Juliet are the eternal [5] symbol of love.

In the play *Hamlet* Ophelia sang this song:

1. **gallant (beau)** : 華美的。
2. **mate** : 交配。
3. **conquerors** : 征服者。
4. **combined** : 結合。
5. **eternal** : 永恒的。

British and American Festivities

'Tomorrow is
St. Valentine's Day
All in the morning betime,
And I a maid[1] at your
window
To be your Valentine.'

Laurence Olivier as Hamlet and Jean Simmons as Ophelia in Olivier's 1948 film.

In 17th century London, sweethearts exchanged presents on 14 February. The English settlers [2] took this romantic tradition to the New World.

Some settlers made beautiful Valentine's cards by hand. They painted butterflies [3], flowers, cupids and hearts on the cards, and then wrote original verses [4].

In the 1800's few people had time to make Valentine's cards. American manufacturers [5] printed millions of romantic cards. The most popular cards had moving [6] parts: windows that opened and showed a romantic poem, and little birds that seemed to fly.

1. **maid**：女孩。
2. **settlers**：定居者。
3. **butterflies**：
4. **verses**：詩句。
5. **manufacturers**：製造商。
6. **moving**：感人的。

Valentine's Day

After the romantic Valentine, the comic [1] Valentine became popular. It had funny cartoons or grotesque [2] pictures with comic messages.

Today there are all types of Valentine's cards: romantic, poetic, comic, grotesque, artistic [3] and handmade [4]!

In 1947 the town of Loveland in Colorado, USA, became Cupid's [5] residence [6]! In Loveland something very unusual happens about two weeks before Valentine's Day. The Loveland Post Office receives more than 300,000 Valentine's cards from all over the world.

The Loveland Post Office cancels [7] these cards with the Loveland,

1. **comic** : 可笑的。
2. **grotesque** : 怪異的。
3. **artistic** : 精美的。
4. **handmade** : 手工製作的。
5. **Cupid** :
6. **residence** : 住處。
7. **cancels** : 蓋郵戳。

77

British and American Festivities

Colorado cancellation[1]. A picture of Cupid and romantic verses are stamped[2] on the envelope[3]. Then the Valentine's cards are sent to the addressee[4].

These are two examples of the Loveland verses:

'Across the land
we send hugs and kisses.
From the Sweetheart City
Come Valentine wishes.'

'Hope this special day
will make you smile
and the love sent from Loveland
will shorten each mile.'

If you want to use this unusual Valentine's service, write to the following address for more information:

Ms Julie Farnham – Visitors Center – Valentine's Cards
5400 Stone Creek Circle - Suite 100 – Loveland, Colorado 80538 - USA
Telephone: 001-880-258-1278

In Great Britain it is customary[5] to send an anonymous[6] Valentine's card or present to the person you like or love. The person must guess who wrote it!

Today in the United States Valentine's Day has a bigger meaning. On this day you can send a card or a present to a parent, grandparent, relative or a friend. Valentine's Day is an occasion to express love or affection[7] to anyone.

1. **cancellation :**

2. **stamped :** 蓋章。

3. **envelope :** 信封。

4. **addressee :** 收卡人。

5. **customary :** 習慣的。

6. **anonymous :** 匿名的。

7. **affection :** 喜愛。

Valentine's Day

In elementary school American boys and girls write cards to their favourite classmates. Then they put them into a big box in the school hall. In the afternoon there is a Valentine's Day party. The teacher takes the cards out of the box and gives them to the children. This is always an exciting moment! Children like counting their cards. Most children sign their Valentine's card, but some remain anonymous or write 'From a Secret Admirer.'

In American high schools and universities there is a 'Sweethearts' Ball' on the evening of 14 February. For the 'Sweethearts' Ball' the hall is decorated with red hearts and cupids! Some people have parties at home to celebrate this day dedicated to love.

In the USA, the top of the Empire [1] State Building in New York City is illuminated with red lights to celebrate Valentine's Day.

The Empire State Building.

1. **Empire** : 帝國。

UNDERSTANDING THE TEXT

 Fill in the gaps with the correct word from the heart.

presents   settlers

Hamlet

anyone   printed

love   Roman

mystery   unusual

cards

a. Sweethearts celebrate Valentine's Day with .... and ....

b. Lupercalia was a .... festival celebrated on 15 February.

c. The origin of the name Valentine is still a ....

d. Valentine's Day is mentioned in Shakespeare's play ....

e. The English .... took the Valentine tradition to the New World.

f. In the 1800's American manufacturers .... millions of Valentine's cards.

g. The town of Loveland in Colorado, offers a very .... service.

h. In the United States, Valentine's Day is an occasion to express .... or affection to ....

2 **Use the prepositions** (介詞) **in the Cupid to complete the following sentences.**

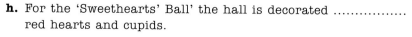

a. Every young man took the name of a young lady .... an urn.

b. A legend says that birds begin to mate .... 14 February!

c. The English settlers took the romantic tradition .... the New World.

d. A picture .... Cupid and romantic verses are stamped on the envelope.

e. The children put the cards .... a big box.

f. There is a party .... the afternoon.

g. The teacher takes the cards .... of the box.

h. For the 'Sweethearts' Ball' the hall is decorated .... red hearts and cupids.

out
on
from
to
with
of
in
into

3 **Listen from page 78 'In Great Britain...' to page 79 '... Secret Admirer' and circle the words you hear.**

give   send   people   person   love   hate   guess   say

car   card   grandmother   write   post   box   class

party   disco   children   child   mother   secret

Cupid's Love Tea

Make your Valentine's Day a special one. Drink a cup of Cupid's love tea with your sweetheart!

Ingredients

4 cups of cold water

1 cup of apple juice

1/4 cup of honey

1/4 teaspoon of cinnamon powder

3 Rosehip tea bags

Boil 1 cup of water and the juice, then add the honey and the cinnamon powder. Stir [1] well. Add the tea bags and let them stand for an hour. Then add 3 cups of cold water. Serve in tea cups or pour into tall glasses with ice cubes [2]. Have a happy Valentine's Day!

1. **stir**：攪動。

2. **cubes**：立方體。

CHAPTER TEN

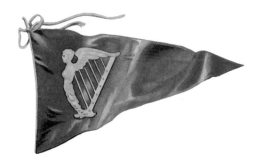

St. Patrick's Day

In the United States and Great Britain, the Irish communities celebrate St. Patrick's Day on 17 March.

The Irish community in the United States is very big. During the 1800's thousands of Irish immigrants went to America. They settled in all parts of the country and took their traditions with them.

Who was St. Patrick and why is he important to the Irish?

St. Patrick is the patron[1] of Ireland. He left several writings about his life and work in Ireland.

Patrick was born in Britain in 389 AD. His father was a landowner[2] and a Christian. When Patrick was 16 years old he was captured and taken to Ireland. He became a slave. After six years Patrick escaped to France by ship. There he became a priest[3].

He returned to Ireland as a bishop[4] in 432 AD. His missionary[5] work

1. **patron** : 保護神。
2. **landowner** : 地主。
3. **priest** : 神職人員。

4. **bishop** : (天主教的) 主教。
5. **missionary** : 傳教的。

British and American Festivities

was very successful. He converted [1] most of the Irish to Christianity. He was a generous and dedicated [2] man. He introduced the Roman alphabet [3] to Ireland.

There are legends about St. Patrick. One legend says that he banished [4] all the serpents [5] from Ireland. Another legend says that he

Irish people celebrating St. Patrick's Day in Dublin, Ireland.

1. **converted**：轉變。
2. **dedicated**：有獻身精神的。
3. **the Roman alphabet**：羅馬字母。
4. **banished**：驅逐。
5. **serpents**：毒蛇。

St. Patrick's Day

used a shamrock [1] to teach the Irish about the Trinity [2]. Today the shamrock is still a symbol of Ireland and the Irish.

St. Patrick died in 461 and his tomb at Downpatrick is a centre of pilgrimage [3].

Today on 17 March there are big parades in most American cities in honour of [4] the Irish. Marching bands dressed in green play lively [5] Irish music. The Irish community organises parades and other colourful events.

American and British people wear green clothing on St. Patrick's Day. Most people wear a green shamrock. Many shops and restaurants are decorated with green shamrocks in honour of the Irish.

A St. Patrick's Day parade in New York.

1. **shamrock**：三葉草 (愛爾蘭的國花)。

2. **Trinity**：(基督教) 三位一體 (即聖父、聖子、聖靈合成一神)。

3. **pilgrimage**：朝聖。

4. **in honour of**：紀念。

5. **lively**：活躍的。

UNDERSTANDING THE TEXT

1 Are the following sentences true (T) or false (F)? Correct the false ones.

	T	F
a. On 17 March, the Irish communities in the United States and Great Britain celebrate St. Patrick's Day.	☐	☐
b. The Irish community in the United States is very small.	☐	☐
c. St. Patrick is the patron of Ireland.	☐	☐
d. When Patrick was 16 years old he was captured and taken to France.	☐	☐
e. When he was a bishop he converted most of the Irish to Christianity.	☐	☐
f. The serpent is a symbol of Ireland and the Irish.	☐	☐
g. American and British people wear green clothing on St. Patrick's Day.	☐	☐

2 How many words can you find in the word river? Circle them in red.

colourIrishizparadeskbyalphabetofgshamrockopslaveitthegenerousrcisbyserpentsj

Use some of the circled words to fill in the gaps below and complete the sentence.

Green _ _ _h_ _ _1_ _ _ _ _ the _r_ _ _ _ .

In Chapter 10 we used a nationality adjective（表示國籍的形容詞）.
Look at this example:

In the 1800's thousands of *Irish* immigrants went to America.

Irish = from Ireland.

Use the words in the shamrock to complete these sentences.

French
Chinese
German
American   British   Dutch
Italian   Celtic   Spanish

a. Patrick was born in Britain. He was ....

b. She lives in the United States. She is ....

c. Patrick escaped to France and lived with the ....

d. Christopher Columbus was born in Italy.
He was ....

e. The .... King and Queen gave Columbus three ships.

f. She comes from China and celebrates the .... New
Year.

g. Halloween has ancient .... origins.

h. The .... Prince Albert, Queen Victoria's husband,
took the Christmas tree to the British Royal Family.

i. The tradition of Santa Claus comes from Holland. The
.... settlers took it to America.

CHAPTER ELEVEN

Easter [1]

aster is a very important Christian festivity. People celebrate the resurrection [2] of Jesus Christ on the third day after his death.

Easter is celebrated on the first Sunday after the full moon in March. This is between 22 March and 25 April.

The origin of the English word Easter is uncertain [3]. It probably comes from the pagan goddess [4] of spring, Eostre. The pagans celebrated the arrival of spring with special festivals.

The Christian Easter slowly replaced [5] the pagan festivals, but some symbols of the spring festivals remained, such as flowers, eggs, rabbits and bunnies [6]. The tradition of the Easter basket is ancient. The pagans offered their eggs in grass baskets to the goddess Eostre.

In Britain and in the United States it is usual to send Easter cards to friends and relatives. For most Christian families Easter morning begins with a church service. Some Easter services begin very early in

---

1. **Easter**：復活節。
2. **resurrection**：復活。
3. **uncertain**：不肯定的。
4. **goddess**：女神。
5. **replaced**：取代。
6. **bunnies**：

Easter

the morning. Others take place [1] out of doors in a garden or park.

Easter is a time for Christians to be happy and there is special music in the churches. It is possible to hear George Frederick Handel's 'Messiah' during some Easter services.

To celebrate Easter and the arrival of spring almost everyone wears something new: a new dress, a new suit, new shoes.

Before Easter Day American and British children paint Easter eggs with bright colours and designs. Coloured eggs were exchanged at ancient spring festivals. The egg is an ancient symbol of life and fertility [2]. The decoration of Easter eggs began in England during the Middle Ages. Members of noble families gave one another gold-covered eggs as Easter presents!

On Easter Day in the United States there is the traditional Easter egg hunt in every city and town. Parents tell their children that the Easter bunny hid [3] many eggs in the park. The children

An egg tree.

must find the hidden eggs. Eggs are hidden in the grass, in a shrub [4] or under a tree. When the children find the eggs they put them in colourful Easter baskets. The child with the biggest number of eggs is the winner.

1. **take place**：發生。
2. **fertility**：生育力。

3. **hid**：動詞 hide 的過去時。

4. **shrub**：

89

British and American Festivities

The tradition of the Easter bunny and the basket of eggs was introduced by German immigrants long ago.

On Easter Monday children roll [1] their eggs down a hill. Egg rolling is an ancient Easter tradition. The first egg that reaches the bottom of the hill without breaking is the winner. A famous egg-rolling contest takes place outside the White House in Washington DC on Easter Monday. The President of the United States invites children to roll eggs on the lawn [2] of the White House.

At Easter time there are chocolate eggs, chocolate bunnies and sugar eggs in all sweet shops. Some eggs are personalized with [3] a name on them.

Hot Cross Buns are a special Easter food. They are small sweet cakes with a cross [4] on top. This cross represents Christ's death on the cross. The buns are usually eaten in Britain on Good Friday.

In many American cities and towns there is an Easter Bonnet [5] Parade. Girls and women with funny and bizarre [6] Easter bonnets march in the parade. Most women make their own bonnets. There is a prize for the most original. The most famous Easter Bonnet Parade is in New York City, on Fifth Avenue. Thousands of people participate[7]!

1. **roll** : 滾動。
2. **lawn** : 草坪。
3. **are personalized with** : 標明。
4. **cross** : 十字架。
5. **Bonnet** : 女帽。
6. **bizarre** : 怪誕的。
7. **participate** : 參加。

Easter Parade in New York.

British and American Festivities

In 1933 the great American composer [1] Irving Berlin wrote a song about the Easter Bonnet Parade. Here is a verse:

'In your Easter bonnet
With all the frills [2] upon it
You'll be the grandest lady
In the Easter parade.'

Families usually spend Easter Day together. The traditional Easter meal consists of [3] roast lamb, peas, new potatoes and eggs in many forms [4].

Participating in an Easter Bonnet Parade.

1. **composer** : 作曲家。
2. **frills** : 飾邊。
3. **consists of** : 包括。
4. **forms** : 形式。

UNDERSTANDING THE TEXT

 Choose the correct answer.

a. Easter is celebrated
- ☐ after the moon in March
- ☐ on the first Sunday in April
- ☐ between 22 March and 25 April

b. The word Easter probably comes from
- ☐ the pagan goddess of spring, Eostre
- ☐ an Anglo-Saxon song
- ☐ a pagan religion

c. Easter baskets, flowers, eggs and rabbits are
- ☐ Christian symbols
- ☐ symbols of pagan spring festivals
- ☐ symbols of an egg rolling contest

d. The decoration of Easter eggs began
- ☐ during a pagan festival
- ☐ in Germany long ago
- ☐ in England during the Middle Ages

e. On Easter Day in the United States there is a
- ☐ traditional Easter egg hunt in every city and town
- ☐ big picnic in the park
- ☐ concert of special music in the park

f. In New York City thousands of people participate in the
- ☐ egg-rolling contest
- ☐ big egg hunt
- ☐ Easter Bonnet Parade

Find the wrong word.

Circle the wrong word in each sentence and write the correct word in the egg. You will find the correct words in the rabbit.

Christian   roll
Bonnet   spring
before   cards

a. Easter is a pagan festivity.

b. The pagan goddess of summer was Eostre.

c. In Britain and in the United States it is usual to send Easter presents.

d. After Easter Day American and British children paint Easter eggs.

e. The President of the United States invites children to find eggs on the lawn of the White House.

f. In many American cities and towns there is an Easter Egg Parade.

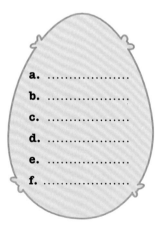

a. ....
b. ....
c. ....
d. ....
e. ....
f. ....

3 You come from Great Britain and you are visiting your cousin in New York City. You want to write a letter to your best friend in Britain telling him/her about your Easter holiday. Put the verbs （動詞）in the Past Simple tense （過去時） and add the articles （冠詞）, if necessary.

Dear .... ,

I / arrive in / New York City / last Friday / at / 6pm.

My cousin / meet / me / at / airport.

We / go / to his home / and / have / dinner. Then we / paint / Easter eggs/ and / eat / ice cream.

On Easter Day / there / be / big / Easter Egg Hunt / in / park.

I / find / fifteen eggs / and / I / win / first prize!

Later / we / go / to see / Easter Bonnet Parade. It / be / fun!

We / see / a lot of / funny / bonnets. One girl / have / toy / aeroplane / on / her bonnet!

In / evening / we / go / to church.

From

....

4 Do you paint Easter eggs?
What is the traditional Easter food in your country?

CHAPTER TWELVE

May Day [1]

On the first day of May we celebrate the arrival of spring and warm weather. Everyone is happy to say goodbye to winter and to welcome spring. Flowers bloom [2], leaves grow and baby animals are born in spring.

The May Day festival originated in the Roman Empire in about 258 BC. It was a festivity in honour of Flora, the goddess of flowers and spring. When the Romans invaded [3] Britain the festivity was celebrated by the Britons too.

The ancient Celts celebrated spring the night before May Day. This festival was called Beltane. The Druids made big fires on top of the hills. The Celts sat around the fires and ate, sang songs and danced happily.

In central Scotland Beltane is still an important festival.

In the Middle Ages May Day was one of the merriest [4] festivities in Britain. People went into the forest after midnight to pick flowers and plants. They returned the next morning and put the leaves and flowers in their hair. There was music, singing and dancing all day.

---

1. **May Day** : 五朔節。
2. **bloom** : 開花。
3. **invaded** : 侵略。
4. **merriest** : 最歡樂的。

May Day

The most important event of the day was the dance around the Maypole. The most beautiful young woman of the village became the Queen of the May. There were Morris dances too.

Dancing around the Maypole.

In Britain May Day is a public holiday and it is celebrated on the first Monday in May. It is not a public holiday in the United States.

In Britain and the United States there are still traditional Maypole dances on May Day. People in costumes dance around the Maypole. The Maypole is decorated with many coloured ribbons. The coloured ribbons represent the sun's rays [1] and form a design on the Maypole.

In Britain many people in costumes do Morris dancing on May Day. They dance with bells and handkerchiefs [2]. The tradition probably came

---

1. **rays** : 光線。
2. **bells and handkerchiefs** :

British and American Festivities

from Spain in the 13th century when it was called Moorish dancing.

In the United States most elementary schools celebrate May Day with dances around the Maypole. In some American towns and universities there are May Day festivals. A young woman becomes the May Queen. Music, dancing around the Maypole and games are all part of the festival. There are beautiful floral [1] decorations everywhere.

Morris dancers.

1. **floral** : 花的。

May Day

Traditional Morris dancing.

UNDERSTANDING THE TEXT

 Match the two parts of the sentences.

a. On the first day of May	**1.** Morris dancing.
b. The May Day festival originated	**2.** we celebrate the arrival of spring.
c. The ancient Celts celebrated spring	**3.** one of the merriest festivities in Britain.
d. In the Middle Ages May Day was	**4.** with a festival called Beltane.
e. People in costumes	**5.** in the Roman Empire in about 258 BC.
f. In Britain people in costumes do	**6.** dance around the Maypole.

 Adverbs （副詞）

Look at this sentence from Chapter 12:

The Celts sat around the fires and ate, sang songs and danced *happily*.

Happily **is an adverb. Adverbs describe verbs** （動詞）**, adjectives** （形容詞） **or other adverbs. Adverbs answer questions such as: how? when? where?**

Look at these examples:

He ate *quickly*.
quickly **describes how he ate**

They danced around the Maypole *today*.
today **describes when they danced**

We celebrate May Day *here*.
here **describes where we celebrate May Day**

Fill in the gaps using the adverbs in the basket.

yesterday

tomorrow   there

loudly   never

today   slowly

a. .... was Beltane and .... is May Day.

b. The Druid said, 'Let's make a fire ...., on top of the hill!'

c. He .... dances with the Morris dancers.

d. On May Day everyone sings ....

e. The children will pick the flowers .... morning.

f. They were tired and they danced ....

3 **Is May Day a holiday in your country?**
If so, what happens?

CHAPTER THIRTEEN

Independence Day

The most animated [1] American festivity is the Fourth of July or Independence Day. It is the nation's birthday. The Fourth of July is a salute [2] to freedom and democracy. It is a time to remember America's ideals of liberty [3], equality and opportunity for all.

What happened on 4 July?

In the 1700's the thirteen American colonies belonged to Great Britain. The colonists did not want to be governed by Britain. They wanted to be independent and to choose their own government. They wanted a democracy. Britain imposed [4] high taxes and there were many rebellions [5].

On 4 July, 1776 a group of patriots [6] wrote the Declaration of Independence [7]. The Declaration proclaimed [8] independence from Britain and democracy and justice for all. With the Declaration of Independence the thirteen colonies created their own nation, the

1. **animated** : 活躍的。
2. **salute** : 敬禮。
3. **liberty** : 自由。
4. **imposed** : 徵收。
5. **rebellions** : 反抗。
6. **patriots** : 愛國者。
7. **Declaration of Independence** : 獨立宣言。
8. **proclaimed** : 宣告。

Independence Day

United States of America. The Liberty Bell is a symbol of Independence Day.

The Americans fought [1] against the British in the American Revolution. After years of war the British were defeated [2] in 1781 in Yorktown, Virginia.

Today Americans celebrate the Fourth of July in many different ways. There is an American flag on every flagpole [3] and many people put a flag outside their window. Americans call their flag 'the Stars and Stripes [4]'.

A U.S. Air Force salute on 4th July.

1. **fought** : 動詞 fight 的過去時。
2. **were defeated** : 被打敗。
3. **flagpole** :
4. **Stars and Stripes** :

British and American Festivities

Every city and town organises its own celebration. Red, white and blue decorations fill the streets.

Traditional Fourth of July events are patriotic speeches, parades, baseball games, competitions, a lot of music, dancing and picnics. These picnics are an old American tradition. The typical picnic consists of hamburgers [1], hot dogs [2], potato salad, chocolate cake and ice cream. The festivities usually end with a brilliant fireworks show [3].

In the West rodeos are a favourite event. A western rodeo is a spectacular event to watch!

Fourth of July umbrellas.

1. **hamburgers** : 漢堡包。
2. **hot dogs** : 熱狗。

3. **fireworks show** : 煙花表演。

A brilliant fireworks show!

British and American Festivities

In Virginia there are historic parades with people in 18th-century costumes.

In New York City the top of the Empire State Building is illuminated with red, white and blue lights!

In Flagstaff, Arizona, American Indians celebrate with a three-day pow-wow [1], a rodeo and tribal dances.

A rodeo.

At a pow-wow.

Bridgeport, California is a small town in the Sierra Nevada Mountains. Bridgeport celebrates Independence Day in an old-fashioned [2] way.

Before the 10 o'clock parade someone reads the Declaration of Independence to the town. Then cowboys and Indians from nearby ranches [3] come to the town on their beautiful horses. Children ride decorated bicycles in the parade. There is a big pie-eating competition. After a delicious picnic of barbecued meat [4], there is a baseball game. In the evening there is country music and dancing.

---

1. **pow-wow** : 議事會。
2. **old-fashioned** : 老式的。
3. **ranches** : 牧場。
4. **barbecued meat** : 烤肉。

UNDERSTANDING THE TEXT

1 **Choose the correct answer.**

a. Independence Day
- [ ] was the first day of the American Revolution
- [ ] is America's birthday
- [ ] is an American Indian festivity

b. In the 1700's the thirteen American colonies
- [ ] were independent
- [ ] declared war against France
- [ ] belonged to Great Britain

c. On 4 July, 1776, a group of patriots
- [ ] wrote the Declaration of Independence
- [ ] started the American Revolution
- [ ] won the war against Britain

d. The Declaration of Independence
- [ ] imposed high taxes on the colonists
- [ ] was a declaration of war
- [ ] proclaimed independence from Britain and democracy and justice for all

e. Rodeos are a favourite Fourth of July event in
- [ ] the West
- [ ] New York City
- [ ] Yorktown, Virginia

f. American Indians celebrate with a three-day
- [ ] picnic and fireworks
- [ ] pow-wow, a rodeo and tribal dances
- [ ] parade and a marching band

Have fun with this crossword puzzle!

Across

1.

2. a symbol of Independence Day

3.

4. an American state
5. there were .... colonies
6. rodeos are a favourite event in the ....

Down

7. colours of the American flag: red, white and ....

8.

9. the British were defeated in ....

10. the colonies belonged to ....

11.

12.

Where is the celebration?

Independence Day is celebrated in different ways in the United States. Match the picture with the name of the place.

a. Arizona

b. Virginia

c. The West

d. New York

 2

 1

 3

4

4 Listen to Chapter 13 and put the pictures in the order they are mentioned.

a ☐

b ☐

c ☐

d ☐

e ☐

g ☐

f ☐

Declaration of Independence

Before the o'clock parade a person read
the Declaration of Independence in the town.
Then cowboys and Indians from nearby
ranches come to town on their beautiful
horses. Children ride decorated bicycles
in the parade.

July 4, 1776

h ☐

Notting Hill Carnival [1]

The Notting Hill Carnival of London is the second biggest carnival in the world and the biggest street festival in Europe! It is always on the last Sunday and Monday in August which is a Bank Holiday (a public holiday) in the United Kingdom. On Sunday there is the Children's Carnival. The Carnival takes place in Notting Hill, West London. When did the first Carnival take place?

In the 1950's people from the Caribbean, and in particular [2] from Trinidad, emigrated [3] to Britain. They took their customs and traditions with them. People remembered the great Carnivals held in the West Indies and in 1964 a street festival took place in Notting Hill.

There were few people in costume dancing in the streets and carrying steel [4] drums in this first festival. However, it was a great success. Since

---

1. **Notting Hill Carnival** : 諾丁山狂歡節。
2. **in particular** : 特別。
3. **emigrated** : 移居外國。
4. **steel** : 鋼。

British and American Festivities

A float with a steel band.

then Carnival has taken place every year in Notting Hill and it has grown into an enormous multicultural arts festival.

Carnival celebrations normally [1] take place before Easter, in the month of March. However, in Britain the celebrations take place in August when the weather is warmer. During the year the West Indian families prepare their beautiful costumes and practise playing their steel drums. They also work on their floats.

On the days of Carnival, Notting Hill is full of colour, excitement, music, noise and people. About two million people go to Carnival every year!

People with wonderful costumes dance in the streets and steel bands play Calypso, the traditional music of the West Indies. They also play Soca, the traditional music of Carnival, a mixture of Soul [2] and Calypso [3]. It is also possible to hear reggae [4], hip-hop [5] and jazz [6]. A parade with colourful floats travels

Colour and excitement at Carnival.

1. **normally** : 通常地。
2. **soul** : 靈歌。
3. **Calypso** : 即興諷刺歌。
4. **reggae** : 雷蓋樂。
5. **hip-hop** : 説唱音樂。
6. **jazz** : 爵士樂。

Notting Hill Carnival

seven and a half kilometres through the streets of London. There is a prize for the best float. It is sometimes almost impossible for the public to walk along the streets – the only way to move is to dance!

In the streets food vendors [1] sell meat and vegetable patties [2], salted fish and other delicious West Indian specialities [3]. Everyone has a good time!

Colourful Carnival costumes.

1. **vendors** : 小販。
2. **patties** : 餡餅。

3. **specialities** : 特產。

113

UNDERSTANDING THE TEXT

1 **Are the following sentences true (T) or false (F)? Correct the false ones.**

		T	F
a.	The Notting Hill Carnival of London is the biggest carnival in the world.	☐	☐
b.	Notting Hill is in West London.	☐	☐
c.	In the 1950's people from Trinidad emigrated to Britain.	☐	☐
d.	The first street festival was not a great success.	☐	☐
e.	In Britain celebrations take place in January when it is cold.	☐	☐
f.	During the Notting Hill Carnival there are wonderful costumes, traditional music and colourful floats.	☐	☐
g.	About two million people go to Carnival every year!	☐	☐

2 **Circle the words that are part of the Notting Hill Carnival.**

noise   school   boats   dragon   snow

steel drums   church   Calypso   meat

April   costumes   parade   forest

police   salted fish   garden   dance

float   doctor   prize

Now use the circled words to fill in the gaps in the sentences.

a. The .... make a lot of ....

b. The .... travels seven and a half kilometres through the streets of London.

c. There is a .... for the best ....

d. The people eat .... patties and ....

e. Everyone listens to .... music.

f. People wear beautiful .... and .... in the streets.

3 **Look at the nouns**（名詞）**in the drums. Then look at the adjectives**（形容詞）**in the box. Put the adjective under the word that describes it. There can be more than one adjective for each noun.**

> beautiful   delicious   loud   long   short
> Calypso   colourful   Soca

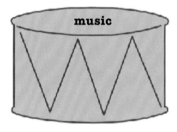

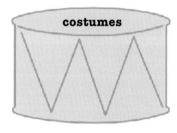

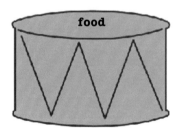

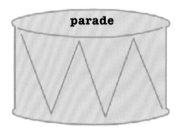

4 **Do you celebrate Carnival? If so, when?**

Internet sites

If you want to know more about the festivities in this book, here are some Internet sites for you.

Halloween :
http://piute.lancaster.k12.ca.us/hcostum.html

Thanksgiving :
http://www.plimoth.org/thanksgi.html

Christmas :
http://www.maui.net/~mcculc/xmas.htm
http://www.santalive.com/family.html

The Pasadena Tournament of Roses :
http://www.citycent.com/sharp

Chinese New Year :
http://www.tat-usa.com/festival/fhistory.htm

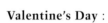

Valentine's Day :
http://www.omn.com/Phone/Valentin.htm
http://www.usacitylink.com/cupid/cletter.html

Easter :
http://www.moray.gov.uk/press/mar98/parade.html

Independence Day :
http://www.co.york.va.us/citizen_news/fourth.htm
http://www.americansbirthday.com

Notting Hill Carnival :
http://www.nottinghillcarnival.net.uk

EXIT TEST

COMPREHENSION

1 Are the following sentences true (T) or false (F)? Correct the false sentences.

		T	F
a.	Columbus Day is an American festivity and it is celebrated on 12 October.	☐	☐
b.	Halloween has ancient Indian origins.	☐	☐
c.	The fifth of December is Guy Fawkes' Night in America.	☐	☐
d.	The Pilgrims were the founders of America. They were the first to celebrate Thanksgiving.	☐	☐
e.	The word Christmas comes from the Roman words 'Cristes maesse', Christ's mass.	☐	☐
f.	The modern Christmas tree originated in the United States.	☐	☐
g.	In Scotland Hogmanay, or New Year's Eve, is the most important celebration of the year.	☐	☐
h.	In California the biggest and oldest New Year's Day events are the 'Tournament of Roses' parade and the Rose Bowl game.	☐	☐
i.	On the Chinese calendar every year has an insect's name.	☐	☐
j.	The custom of celebrating Valentine's Day probably began with the Roman festival of Lupercalia on 15 February.	☐	☐
k.	The Irish community in the United States is very small but it celebrates St. Patrick's Day on 17 March.	☐	☐
l.	The word Easter probably comes from the pagan goddess of spring, Eostre.	☐	☐
m.	In Great Britain many people in costumes do Morris dancing on Easter Day.	☐	☐
n.	On 4 July, 1776 a group of American patriots wrote the Declaration of Independence. Today the fourth of July or Independence Day celebrates America's birthday.	☐	☐

o. The Notting Hill Carnival of London is the biggest street festival in Europe. □ □

GRAMMAR

2 **Fill in the gaps with the correct superlatives. Use the adjectives in brackets.**

a. California is the (*sunny*) .... place in America.

b. The most colourful costume won first prize because it was the (*good*) ....

c. Summer is the (*hot*) .... season of the year.

d. Jimmy was the (*small*) .... child at the park.

e. Your cooking is bad but mine is the (*bad*) ....

f. Susan is the (*pretty*) .... girl in the class.

g. Alice and Edward are the (*happy*) .... couple.

3 **Put the verbs in brackets in the Past Simple tense.**

At Halloween, the children (*take*) [1]................. their costumes to school; in the afternoon they (*put*) [2]................. them on and (*have*) [3]................. a Halloween party. Some children (*wear*) [4]................. scary costumes. The best costume (*win*) [5]................. a prize. The children (*play*) [6]................. games and (*eat*) [7]................. candied apples and pumpkin pie.

In the evening children and teenagers (*go*) [8]................. "trick-or-treating." People (*give*) [9]................. them sweets or money. The children (*be*) [10]................. happy with the treats.

4 **Write the nationality adjective for each country.**

France	....
Spain	....
China	....
Germany	....
the United States of America	....
Great Britain	....
Italy	....
Mexico	....
Holland	....
Ireland	....

5 **What is your favourite festivity and why?**

British and American Festivities

KEY TO THE EXERCISES AND EXIT TEST

CHAPTER 1

Page 12 Exercise 1
a. T
b. F - Christopher Columbus wanted to reach the East by sailing to the West.
c. F - The King and Queen of Spain gave Columbus three ships: the *Santa Maria*, the *Niña* and the *Pinta*.
d. F - Columbus was an Italian navigator and explorer.
e. T
f. F - Most Americans celebrate Columbus Day with colourful parades.
g. T

Page 12 Exercise 2
a. round
b. coast
c. beach
d. horizon

a. coast
b. horizon
c. round
d. beach

Page 13 Exercise 3
Marco Polo, Magellan, Sir Francis Drake, Captain Cook, Columbus

CHAPTER 2

Page 19 Exercise 1
a. the United States and Great Britain
b. Celtic
c. All Hallows' Eve
d. have a Halloween party at school
e. have a party on Halloween evening
f. go 'trick-or-treating'

Page 20 Exercise 2
a. was
b. practised
c. made, dressed
d. believed, came
e. became
f. took
g. gave
h. wore

Page 21 Exercise 3
a. alien
b. devil
c. Batman
d. skeleton
e. witch

Page 22 Exercise 4
children, school, witch, people, buy, afternoon, party, with, make, windows, Celtic

119

CHAPTER 3

Page 28 Exercise 1

a. a British
b. Protestant, Catholics
c. Catholics, blow up
d. guard
e. King's soldiers, were hanged
f. Britain, Guy
g. night

Page 28 Exercise 2

a. Who
b. When
c. Who
d. Where
e. Why
f. Where
g. What

Page 29 Exercise 3

newspaper, shirt, trousers, jumper, hair, shoes

CHAPTER 4

Page 34 Exercise 1

a. T
b. T
c. F – The tradition started with the Pilgrims in America.
d. T
e. F – Their first winter was very cold and they had little food.
f. F – Squanto helped the Pilgrims to grow corn, hunt and live in the wilderness.
g. T
h. T

Page 34 Exercise 2

a. Mayflower-3; **b.** wilderness-1;
c. Pilgrims-5; **d.** harsh-2; **e.** Squanto-6;
f. crops-4; **g.** Plymouth-7.

Page 35 Exercise 3

Dear Grandmother,
I arrived in America one year ago.
The winter was very cold and there was little food.
Many people died.
In the spring some kind Indians helped us.
They became our good friends.
We built small houses and cultivated crops.
The harvest was very good this year.
Yesterday we invited the Indians to a Thanksgiving dinner.
We thanked God for everything.
Love from, ...

CHAPTER 5

Page 47 Exercise 1

a. two pagan festivals
b. the birthday of the sun
c. Yule
d. Germany
e. designed in Britain in 1843
f. St. Nicholas
g. an American cartoonist

CHRISTMAS TRADITIONS

FOOD

Page 48 Exercise 1

a. roast ham
b. roast turkey
c. mince pies
d. fruit cake
e. Christmas cookies
f. Christmas pudding

TREES

Page 52 Exercise 1

a. g, h   **b.** a   **c.** c   **d.** f   **e.** b, d, (e)

Page 52 Exercise 2

Open answer.

CHRISTMAS AROUND THE WORLD

Page 54 Exercise 1

a. Great Britain
b. U.S.A.
c. Germany
d. Mexico
e. Lapland
f. Italy

Page 56 Exercise 2

a. Christmas tree
b. Nativity scene
c. teddy bear
d. ornament
e. rocking horse

f. Santa Claus

CHAPTER 6

Page 60 Exercise 1
a. Hogmanay, important
b. parties, balls
c. favourite, Times Square
d. Big Ben
e. promises
f. office, calendars

Page 61 Exercise 2

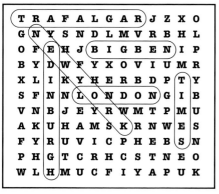

Page 61 Exercise 3
going, party, dance, ring, street, end, make

CHAPTER 7

Page 65 Exercise 1
a. F – People have always celebrated the New Year.
b. T
c. F – In the United States many families have 'Open House' on New Year's Day.
d. T
e. F – The Pasadena "Tournament of Roses" is an important parade.
f. T

Page 65 Exercise 2
a. noisiest
b. sunniest
c. youngest
d. richest
e. coldest
f. best

Page 67 Exercise 3
1f – **2**a – 3e – **4**c – **5**d – **6**b.

CHAPTER 8

Page 71 Exercise 1
a. between 1 January and 19 February
b. has an animal's name
c. wear new clothes with bright colours
d. lucky money
e. Chinatown in San Francisco
f. dragon

Page 72 Exercise 2

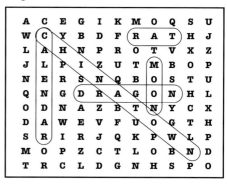

CHAPTER 9

Page 80 Exercise 1
a. cards, presents
b. Roman
c. mystery
d. *Hamlet*
e. settlers
f. printed
g. unusual
h. love, anyone

Page 81 Exercise 2
a. from
b. on
c. to
d. of
e. into
f. in
g. out
h. with

Page 81 Exercise 3
send, person, love, guess, card, write, box, party, children, secret

CHAPTER 10

Page 86 Exercise 1
a. T
b. F – The Irish community in the United States is very big.
c. T
d. F – When Patrick was 16 years old he was captured and taken to Ireland.
e. T
f. F – The shamrock is a symbol of Ireland and the Irish.
g. T

Page 86 Exercise 2
colour, Irish, parades, by, alphabet, of, shamrock, slave, it, the, generous, is, by, serpents

Green is the colour of the Irish.

Page 87 Exercise 3
a. British; **b.** American; **c.** French; **d.** Italian; **e.** Spanish; **f.** Chinese; **g.** Celtic; **h.** German; **i.** Dutch

CHAPTER 11

Page 93 Exercise 1
a. between 22 March and 25 April
b. the pagan goddess of spring, Eostre
c. symbols of pagan spring festivals
d. in England during the Middle Ages
e. traditional Easter egg hunt in every city and town
f. Easter Bonnet Parade

Page 94 Exercise 2
a. Easter is a Christian festivity.
b. The pagan goddess of spring was Eostre.
c. In Britain and in the United States it is usual to send Easter cards.
d. Before Easter Day American and British children paint Easter eggs.
e. The President of the United States invites children to roll eggs on the lawn of the White House.
f. In many American cities and towns there is an Easter Bonnet Parade.

Page 95 Exercise 3
Dear .....,
I arrived in New York City last Friday at 6 p.m.
My cousin met me at the airport. We went to his home and had dinner. Then we painted Easter eggs and ate ice cream.
On Easter Day there was a big Easter Egg Hunt in the park.
I found fifteen eggs and I won first prize!
Later we went to see the Easter Bonnet Parade. It was fun!
We saw a lot of funny bonnets. One girl had a toy aeroplane on her bonnet!
In the evening we went to church.
From ...

CHAPTER 12

Page 100 Exercise 1
a2 - b5 - c4 - d3 - e6 - f1.

Page 100 Exercise 2
a. yesterday, today; **b.** there; **c.** never; **d.** loudly; **e.** tomorrow; **f.** slowly

CHAPTER 13

Page 107 Exercise 1
a. is America's birthday
b. belonged to Great Britain
c. wrote the Declaration of Independence
d. proclaimed independence from Britain and democracy and justice for all
e. the West
f. pow-wow, a rodeo and tribal dances

Page 108 Exercise 2

Page 109 Exercise 3
1c - 2d - 3a - 4b.

Page 110 Exercise 4
a5 - b4 - c6 - d3 - e2 - f1 - g7 - h8.

CHAPTER 14

Page 114 Exercise 1
a. F – The Notting Hill Carnival of London is the second biggest carnival in the world.
b. T
c. T
d. F – The first street festival was a great success.
e. F – In Britain celebrations take place in August when it is warmer.
f. T
g. T

Page 114 Exercise 2
noise, steel drums, Calypso, meat, costumes, parade, salted fish, dance, float, prize

a. steel drums, noise
b. parade
c. prize, float
d. meat, salted fish
e. Calypso
f. costumes, dance

Page 115 Exercise 3
music: beautiful, loud, Calypso, Soca.
costumes: beautiful, long, short, colourful.
food: delicious.
parade: beautiful, loud, long, short, colourful.

1. **a.** F – It is celebrated on the second Monday of October.
   b. F – It has ancient Celtic origins.
   c. F – The fifth of November is Guy Fawkes' night in Britain.
   d. T
   e. F – The word Christmas comes from Old English.
   f. F – It originated in western Germany.
   g. T
   h. T
   i. F – It has an animal's name.
   j. T
   k. F – The Irish community in the United States is very big.
   l. T
   m. F – Many people in costumes do Morris dancing on May Day.
   n. T
   o. T

2. **a.** sunniest **b.** best **c.** hottest **d.** smallest **e.** worst **f.** prettiest
   g. happiest

3. **1.** took **2.** put **3.** had **4.** wore **5.** won **6.** played **7.** ate **8.** went
   9. gave **10.** were

4. France – French
   Spain – Spanish
   China – Chinese
   Germany – German
   The United States of America – American
   Great Britain – British
   Italy – Italian
   Mexico – Mexican
   Holland – Dutch
   Ireland – Irish

5. Open answer.

Notes

 # Notes

Black Cat English Readers

BLACK CAT ENGLISH CLUB
Membership Application Form

BLACK CAT ENGLISH CLUB is for those who love English reading and seek for better English to share and learn with fun together.

Benefits offered: - Membership Card

       - Member badge, poster, bookmark

       - Book discount coupon

       - Black Cat English Reward Scheme

       - English learning e-forum

       - Surprise gift and more...

Simply fill out the application form below and fax it back to 2565 1113.

Join Now! It's FREE exclusively for readers who have purchased *Black Cat English Readers* !

The book(or book set) that you have purchased: _____

English Name: _____ (Surname) _____ (Given Name)

Chinese Name: _____

Address: _____

Tel: _____ Fax: _____

Email: _____

Sex: ❏ Male   ❏ Female   (Login password for e-forum will be sent to this email address.)

Education Background: ❏ Primary 1-3   ❏ Primary 4-6   ❏ Junior Secondary Education (F1-3)

       ❏ Senior Secondary Education (F4-5)   ❏ Matriculation

       ❏ College   ❏ University or above

Age: ❏ 6 - 9   ❏ 10 - 12   ❏ 13 - 15   ❏ 16 - 18   ❏ 19 - 24   ❏ 25 - 34

    ❏ 35 - 44   ❏ 45 - 54   ❏ 55 or above

Occupation: ❏ Student   ❏ Teacher   ❏ White Collar   ❏ Blue Collar

       ❏ Professional   ❏ Manager   ❏ Business Owner   ❏ Housewife

       ❏ Others (please specify: _____)

As a member, what would you like **BLACK CAT ENGLISH CLUB** to offer:

    ❏ Member gathering/ party   ❏ English class with native teacher   ❏ English competition

    ❏ Newsletter   ❏ Online sharing   ❏ Book fair

    ❏ Book discount   ❏ Others (please specify: _____)

Other suggestions to **BLACK CAT ENGLISH CLUB**:

_____

Please sign here: _____

(Date: _____)